SPHERICAL TRIGONOMETRY

AFTER THE CESÀRO METHOD

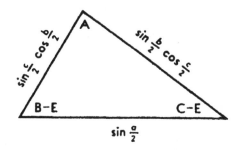

J. D. H. DONNAY, E.M., Ph.D.

INTERSCIENCE PUBLISHERS, INC. - NEW YORK, N. Y.

To the Memory of
GIUSEPPE CESÀRO
1849 – 1939
**CRYSTALLOGRAPHER AND MINERALOGIST
PROFESSOR AT THE UNIVERSITY OF LIÉGE**

PREFACE

From a practical standpoint spherical trigonometry is useful to engineers and geologists, who have to deal with surveying, geodesy, and astronomy; to physicists, chemists, mineralogists, and metallurgists, in their common study of crystallography; to Navy and Aviation officers, in the solution of navigation problems. For some reason, however, spherical trigonometry is not recognized as a regular subject in many American college curricula. As a consequence, the teacher of a science for which a working knowledge of spherical triangles is desirable usually finds he has to impart it to the students himself, or to use ready-made formulae that his listeners have never seen, let alone derived, before.

This book has been written in an attempt to meet this situation. It can be covered in about ten to twelve lectures and could well serve as a text in a one-unit course for a quarter or a semester. It aims at giving the strict minimum, as briefly as possible. The straightforward and time-saving Cesàro method seems particularly suitable for this purpose, tying together as it does, from the outset, the concepts of spherical and plane trigonometry. This method has withstood the test of experience. For years Belgian students have thrived on it. Personally I have taught it for eight years to crystallography students at the Johns Hopkins University, and for two years to freshmen classes at Laval University, with gratifying results.

The order in which the subject matter is arranged may appear unorthodox. It has proved satisfactory, however, from the teaching point of view. Through the use of the stereographic projection (Ch. I), the concept of spherical excess—somewhat of a stumbling block—is mastered from

the start.   A working knowledge of Cesàro's key-triangles is
acquired, as soon as they are established (Ch. II), through
the derivation of Napier's and Delambre's formulae and
several expressions of the spherical excess (Ch. III).   This
much insures the understanding of the method.   The treat-
ment of the oblique-angled triangle (Ch. IV) is thereby so
simplified that one can dispense with that of the right-angled
triangle.   The latter properly follows as a particular case
(Ch. V).   Examples of computations are given (Ch. VI),
both with logarithms and with the calculating machine.
They are followed by a selection of problems, completely
worked out (Ch. VII), and a number of exercises with answers
(Ch. VIII).   Cross-references are made in the theoretical
chapters to appropriate applications.   Proofs believed to be
new are marked by asterisks.

    In conclusion may I be permitted to say that this booklet
was planned jointly by Cesàro and myself, several years ago.
He who was to have been the senior author passed away
shortly afterwards.   In introducing Cesàro's elegant method
to the English-speaking public, I would like to think that
my writing will reflect his influence and, to some degree at
least, his reverence for simplicity and rigor.

J. D. H. D.

Hercules Powder Company Experiment Station
             Wilmington, Delaware
                  April 1945

CONTENTS

SPHERICAL TRIGONOMETRY
AFTER THE CESÀRO METHOD

INTRODUCTION

1. Purpose of Spherical Trigonometry. Spherical trigonometry is essentially concerned with the study of angular relationships that exist, in space, between planes and straight lines intersecting in a common point O. A bundle of planes passed through O intersect one another in a sheaf of straight lines. Two kinds of angles need therefore be considered: angles between lines [1] and angles between planes (dihedral angles).

The spatial angular relationships are more easily visualized on a sphere drawn around O with an arbitrary radius. Any line through O is a diameter, and any plane through O a diametral plane, of such a sphere. The former punctures the sphere in two diametrically opposite points, the latter intersects it along a *great circle*. The angle between two lines is measured on the sphere by an arc of the great circle whose plane is that of the two given lines. The angle between two planes is represented by the angle between the two great circles along which the given planes intersect the sphere. Indeed, by definition, the angle between the great circles is equal to the angle between the tangents to the circles at their point of intersection, but these tangents are both perpendicular to the line of intersection of the two given planes, hence the angle between the tangents is the true dihedral angle.

An open pyramid, that is to say a pyramid without base, whose apex is made the center of a sphere determines a *spherical polygon* on the sphere. The vertices of the polygon are the points where the edges of the pyramid puncture the

---

[1] From now on, the word *line* will be used to designate a straight line, unless otherwise stated.

3

sphere; the sides of the polygon are arcs of the great circles
along which the faces of the pyramid intersect the sphere.
The angles of the polygon are equal to the dihedral angles
between adjacent faces of the pyramid. The sides of the
polygon are arcs that measure the angles of the faces at the
apex of the pyramid, that is to say, angles between adjacent
edges.

A trihedron is an open pyramid with three faces. The
three axes of co-ordinates in solid analytical geometry, for
instance, are the edges of a trihedron, while the three axial
planes are its faces. Consider a trihedron with its apex at
the center of the sphere. It determines a *spherical triangle*
on the sphere.[2] In the general case of an oblique trihedron,
an oblique-angled spherical triangle is obtained, that is to
say, one in which neither any angle nor any side is equal to
90°. The main object of spherical trigonometry is to investi-
gate the relations between the six parts of the spherical triangle, namely its three sides and its three angles.

2. The Spherical Triangle. The sides of a spherical triangle
are arcs of great circles. They can be expressed in angular
units, radians or degrees, since all great circles have the same
radius, equal to that of the sphere. As a consequence of the
conventional construction by means of which the spherical
triangle has been defined (Sn. 1), any side must be smaller
than a semi-circle and, likewise, any angle must be less
than 180°.

By considering the trihedron whose apex is at the center of
the sphere, we see that the sum of any two sides of a spherical
triangle is greater than the third side, that any side is greater
than the difference between the other two sides, and that the
sum of all three sides (called the *perimeter*) is less than 360°.

Because any angle of a spherical triangle is less than 180°,
the sum of all three angles is obviously less than 540°. It is
greater than 180°, as we shall see later (Ch. II, Sn. 1).

[2] This is the "Eulerian" spherical triangle, the only one to be
considered in this book.

3. Comparison between Plane and Spherical Trigonometry.

In plane trigonometry, you draw triangles in a plane. Their sides are segments of straight lines. The shortest path from one point to another is the straight line that connects them. The distance between two points is measured, along the straight line, in units of length. The sum of the angles of a plane triangle is 180°. Through any point in the plane, a straight line can be drawn parallel to a given line in the plane (Euclidian geometry).

In spherical trigonometry, triangles are drawn on a sphere. Their sides are arcs of great circles. The shortest path (on the sphere) from one point to another is the great circle that connects them. The (spherical, or angular) distance between two points is measured, along the great circle, in units of angle. The sum of the angles of a spherical triangle is greater than 180°. Through a point on the sphere, no great circle can be drawn parallel to a given great circle on the sphere (spherical geometry is non-Euclidian).

4. Fields of Usefulness of Spherical Trigonometry.

Most problems dealing with solid angles can be reduced to questions of spherical trigonometry. Such problems crop up in the study of geometrical polyhedra. Problems of solid analytical geometry in which planes and lines pass through the origin usually have trigonometric solutions.

Problems involving spatial directions around one point are encountered in crystallography. Well formed crystals are bounded by plane faces. From a point O taken anywhere inside the crystal, drop a perpendicular on each face; this *face normal* defines the direction of the face. Relationships between the inclinations of the faces relative to one another appear in the network of spherical triangles which the sheaf of face normals determines on a sphere drawn around O.

Surveying is concerned with such small regions of the earth surface that they can be considered plane in a first approximation. Geodesy deals with larger regions, for which the curvature of the earth must be taken into account. In a second approximation the earth is taken as spherical, and the formulae of spherical trigonometry are applicable. (Further refinements introduce corrections for the lack of perfect sphericity of the "geoid.")

In astronomy the application of spherical trigonometry is obvious. The observer occupies a point that is very nearly the center of the celestial sphere around the earth. Each *line of sight* is a radius of the sphere. To the observer who is not aware of, or concerned with, the distances from the earth to the heavenly bodies, the latter appear to move on a sphere. The angle subtended by two stars, as seen by the observer, will thus become a side in a spherical triangle. Navigation techniques, either on the high seas or in the air, being based on astronomical observations, likewise depend on the solution of spherical triangles.

CHAPTER I

THE STEREOGRAPHIC PROJECTION

1. Definition. The problem of representing a sphere on a plane is essentially that of map projections. Of the many types of projections that have been devised, one of the most ancient is the stereographic.

In geographical parlance, used for convenience, the projection plane is the plane of the equator, and the projection point, the South Pole. Points in the Northern Hemisphere are projected inside the equatorial circle; points in the Southern Hemisphere, outside the equatorial circle; any point on the equator is itself its own stereographic projection (Fig. 1). The North Pole is projected in the center of the projection; the South Pole, at infinity.

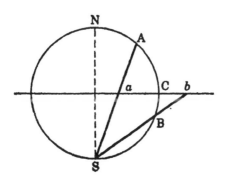

Fig. 1.—The stereographic projection.

2. First Property of the Stereographic Projection. *Circles are projected as circles or straight lines.* If the circle to be projected passes through S, its projection is a straight line. This is obvious, since the projection of the circle is the intersection of two planes: the plane of the circle and the plane of the equator. Note that if the given circle is a great circle passing through S (hence, a meridian) its projection is a diameter of the equator.

If the circle to be projected does not pass through S, its projection is a circle. The proof [1] of this is based on the following theorem.

Consider (Fig. 2) an oblique cone with vertex S and circular base AB. Let the plane of the drawing be a section through S and a diameter AB of the base. The circular base is projected on the drawing as a straight line AB. A section ab, perpendicular to the plane of the drawing, and such that the angles SAB, Sba are equal, is called sub-contrary (or anti-parallel) to the base.

Theorem: In an oblique cone having a circular base, any section sub-contrary to the base is circular.

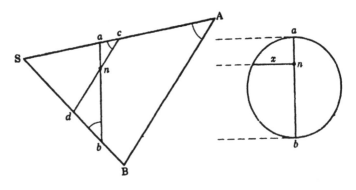

FIG. 2.—Sub-contrary sections.

Take a section cd parallel to the base and, hence, obviously circular. The two sections ab and cd intersect along a common chord, projected at a point n, which bisects the chord. Let x be the length of the semi-chord. The triangles can and bdn are similar (angles equal each to each), hence $an : nd$

[1] This proof can be omitted in a first reading.

$= cn : nb$, or $an.nb = cn.nd$. Because cd is circular, $cn.nd$ $= x^2$. Hence $an.nb = x^2$, and ab is also circular.[2]

Now consider a section of the sphere of projection cut perpendicular to the intersection of the equatorial plane EE' and the plane of the circle AB to be projected (Fig. 3). The angles SAB, Sba (marked on the drawing) are equal, since the measure of SAB = $\frac{1}{2}$(SE' + BE') and that of Sba = $\frac{1}{2}$(SE + BE') are equal (because SE = SE' = 90°). The sections ab and AB of the cone of projection are therefore sub-contrary. Since AB is circular, so is ab.

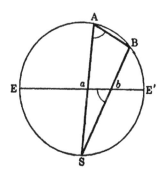

Fig. 3.—Projected circle, a circle.

Remark.—The center of the projected circle is the projection of the vertex of the right cone tangent to the sphere along the given circle. Let C be the vertex of the right cone tangent to the sphere along the given circle AB (Fig. 4). Join CS, intersecting the equator in c and the sphere in D. The angles marked α are equal as having the same measure (one half arc AD). Likewise for the angles marked β, γ, δ. The Law of Sines, applied to the triangles Sac and Scb, gives:

$$ac : Sc = \sin \alpha : \sin \beta, \qquad cb : Sc = \sin \gamma : \sin \delta.$$

[2] This reasoning rests on the theorem, "If, from any point in the circumference, a perpendicular is dropped on a diameter of a circle, the perpendicular is the mean proportional of the segments determined on the diameter," and its converse.

Applied to the triangles ACD and BCD, the same law gives:

$$\sin \alpha : \sin \beta = CD : CA, \qquad \sin \gamma : \sin \delta = CD : CB.$$

Since CA = CB (tangents drawn to the sphere from the same point), these ratios of sines are equal. Whence $ac = cb$, and c is the center of the projected circle.[3]

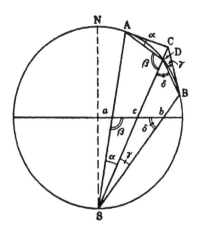

Fig. 4.—Center of projected circle.

Note that if the given circle is a great circle (but not a meridian, nor the equator itself) its projection will be a circle having a radius larger than that of the equator and cutting the equator at the ends of a diameter. A great circle cuts the equator at the ends of a diameter of the latter; points on the equator are themselves their own stereographic projections.

3. Second Property of the Stereographic Projection. *The angle between two circles is projected in true magnitude.*[4] The

---

[3] The proof holds good if the given circle is a great circle; the right cone with circular base becomes a right cylinder with circular base. Make the construction.

[4] A more general property of the stereographic projection is that the angle between any two *curves* on the sphere is projected in true magnitude. The property proved in the text, however, is sufficient for our purpose. This proof can be omitted in a first reading.

angle between two circles drawn on a sphere is equal to the angle between their tangents. We shall prove: (1) that the angle between the projected tangents is equal to the angle between the tangents; (2) that the projected tangents are tangent to the projected circles. Thus will be established the property that the angle between the projected circles is equal to the angle between the circles.

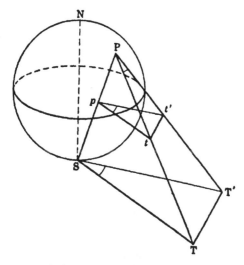

FIG. 5.—Angle between projected tangents equal to angle between tangents.

(1) Consider (Fig. 5) the point P in which two given circles intersect. Let PT and PT' be the tangents to these circles; they cut the plane of the equator in t, t' and the plane tangent to the sphere at S in T, T'. Join PS, intersecting the plane of the equator in p, the stereographic pole of P. The pro-jected tangents are pt, pt'. Join ST and ST'.

The triangles TPT' and TST' are similar (TT' common; TP = TS and T'P = T'S, as tangents drawn to the sphere from the same point). Hence, angle TPT' = angle TST'. Again, the triangles tpt' and TST' are similar (all sides parallel each to each; two parallel planes being intersected by any

third plane along parallel lines).   Hence, angle *tpt′* = angle
TST′.   It follows that the angle between the tangents (TPT′)
and the angle between the projected tangents (*tpt′*) are equal.

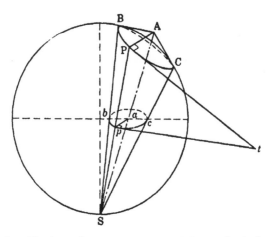

Fɪɢ. 6.—Projected tangent, tangent to projected circle.

*(2) Consider (Fig. 6) a right cone tangent to the sphere
along the given circle PBC; let A be the vertex of this cone.
We know that the circle PBC is projected as a circle *pbc*.
We have seen that the projection of the vertex A is the center
a of the projected circle.   A generatrix AP of the right cone
is projected as a radius *ap* of the projected circle.   P*t*, the
tangent to the given circle at P, is projected in *pt* (*t* in the
plane of the equator).   But the angle *apt*, being the angle
between the projections of the tangents PA and P*t*, is equal
to the angle between the tangents themselves, that is to say
90°.   Hence, *pt* is tangent to the projected circle.

* Proofs believed to be new are marked by asterisks.

CHAPTER II

CESÀRO'S KEY-TRIANGLES[1]

1. Cesàro's "Triangle of Elements." Consider a spherical triangle ABC. Without loss of generality, we may suppose that one of its vertices A is located at the North pole of the sphere. Project this triangle stereographically. The projected triangle A'B'C' (Fig. 7) will have a vertex A' at the

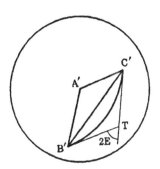

Fig. 7.—Stereographic projection.

center of the projection, and two of its sides, A'B' and A'C', being projected meridians, will be straight lines; its third side B'C' will be an arc of a circle.

At B' and C' draw the tangents to the circle B'C', meeting in T. Because the stereographic projection is angle-true,

$$C'A'B' = A, \qquad A'B'T = B, \qquad A'C'T = C.$$

[1] Cesàro, G. Nouvelle méthode pour l'établissement des formules de la trigonométrie sphérique. *Bull. Acad. roy. de Belgique* (Cl. des Sc.), 1905, 434.
——— Les formules de la trigonométrie sphérique déduites de la projection stéréographique du triangle. Emploi de cette projection dans les recherches sur la sphère. *Bull. Acad. roy. de Belgique* (Cl. des Sc.), 1905, 560.

Designate by 2E the external angle between the tangents meeting in T. It is easy to see that

$$A + B + C - 180° = 2E.$$

The angle 2E is called the *spherical excess* of the spherical triangle ABC. It is equal to the excess over 180° of the sum of the angles of the spherical triangle. We shall express it in degrees.

The angles of the plane triangle A'B'C' are expressed as follows, in terms of the angles of the spherical triangle ABC and its spherical excess 2E,

$$A' = A, \qquad B' = B - E, \qquad C' = C - E.$$

The sides of A'B'C' are functions of the sides of the spherical triangle ABC (Fig. 8). Taking the radius of the sphere

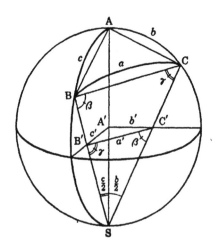

Fig. 8.—Perspective drawing of the sphere of projection.

as the unit of length, we have

$$c' = \tan \frac{c}{2}, \qquad b' = \tan \frac{b}{2}.$$

Each of the quadrilaterals ABB'A' and ACC'A' has two opposite angles equal to 90° (one at A', by construction; the opposite one, as being inscribed in a semi-circle) and is, therefore, inscribable in a circle. Hence

$$SB.SB' = SA.SA' = SC.SC'$$

and BCC'B' are also concyclic. It follows that the angles

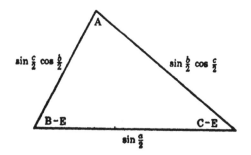

FIG. 9.—Triangle of elements relative to the angle A.

marked β (and the angles marked γ) are equal, so that the triangles SBC and SC'B' are similar.

We may write, therefore,

$$B'C' : BC = SC' : SB,$$

or [2]

$$\frac{a'}{2 \sin \frac{a}{2}} = \frac{\sec \frac{b}{2}}{2 \cos \frac{c}{2}},$$

whence

$$a' = \frac{\sin \frac{a}{2}}{\cos \frac{b}{2} \cos \frac{c}{2}}.$$

[2] For those who prefer step-by-step derivations: $B'C' = a'$, by definition. $BC = 2 \sin \frac{1}{2}a$, for the chord BC subtends an arc a and the chord is equal to twice the sine of half the angle. SC', in the right-angled triangle SC'A', where $SA' = 1$, is the secant of the angle A'SC'. Finally, in the triangle SAB, where the angle at B is a right angle, $SB = SA \cos ASB = 2 \cos \frac{1}{2}c$.

The triangle obtained by multiplying the three sides of A'B'C' by

$$\cos\frac{b}{2}\cos\frac{c}{2}$$

is Cesàro's *triangle of elements* relative to the angle A (Fig. 9).

Other "triangles of elements," relative to the angles B and C, can be obtained by cyclic permutations.

2. Cesàro's "Derived Triangle."   A *lune* is the spherical surface bounded by two great semi-circles; for instance (Fig. 8),

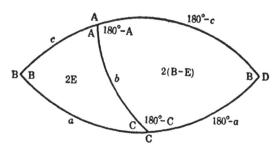

FIG. 10.—The two complementary trihedra, each showing its six parts and spherical excess.

the area ABSCA between two meridians, ABS and ACS. Two trihedra are called *complementary* when the two spherical triangles they determine on the sphere form a lune.   For instance (Fig. 8), the trihedra A'ABC and A'SBC are complementary.   They have two edges, A'B and A'C, in common and the third edge, A'S, of one is the prolongation of the third edge, AA', of the other; the vertices A and S lie at the ends of a diameter, and the two spherical triangles ABC and SBC are seen to form a lune.

Designating by O the center of the sphere, consider a trihedron OABC, represented (Fig. 10) by its spherical triangle ABC.   Produce the great circles BA and BC till they meet, in D, thus forming a lune.   The spherical triangle ADC represents the *complementary trihedron* OADC.   The parts of the triangle ADC are easily expressed in terms of those of the triangle ABC. One side, b, is common; the angle at D is equal to B; the other parts are the supplements of corresponding parts of the triangle ABC (Fig. 10). The spherical excess is found to be $(180° − A) + (180° − C) + B − 180°$ $= 180° − (A + B + C) + 2B = 2B − 2E = 2(B − E)$.

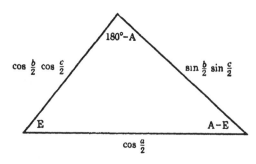

FIG. 11.—Derived triangle relative to the angle A.

Let us compose the triangle of elements, relative to the angle $(180° − A)$, of the complementary trihedron OADC. Its six parts are tabulated below, together with those of the triangle of elements, relative to A, of the primitive trihedron OABC.

THE SIX PARTS OF THE TRIANGLE OF ELEMENTS

For trihedron OABC		For trihedron OADC	
Angles	Sides	Angles	Sides
A	$\sin \dfrac{a}{2}$	$(180° − A)$	$\sin \dfrac{180° − a}{2}$
B − E	$\sin \dfrac{b}{2} \cos \dfrac{c}{2}$	$B − (B − E)$	$\sin \dfrac{b}{2} \cos \dfrac{180° − c}{2}$
C − E	$\sin \dfrac{c}{2} \cos \dfrac{b}{2}$	$(180° − C) − (B − E)$	$\sin \dfrac{180° − c}{2} \cos \dfrac{b}{2}$

The expression of the third angle is transformed as follows:
$180° - C - B + E = A + 180° - (A + B + C) + E = A - 2E + E = A - E$. The other parts are easily simplified.

The new key-triangle can now be drawn (Fig. 11); it is Cesàro's *derived triangle* relative to the angle A. The derived triangle of a trihedron is the triangle of elements of the complementary trihedron.

Other "derived triangles," relative to the angles B and C, can be obtained by cyclic permutations.

3. Polar Triangles. Consider a spherical triangle ABC and the corresponding trihedron OABC. Erect OA*, OB*, OC*, perpendicular to the faces of the trihedron, OBC, OCA, OAB, and on the same sides of these faces as OA, OB, OC, respectively. The new trihedron OA*B*C* determines, on the sphere (Fig. 12), a triangle A*B*C*. The vertices, A*, B*, C*, are called the *poles* of the planes OBC, OCA, OAB, respectively.

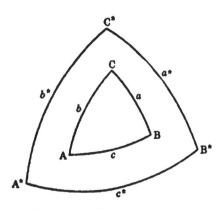

Fig. 12.—Polar triangles.

The triangle A*B*C* is said to be the *polar triangle* of ABC. It follows from the construction that ABC is the polar triangle of A*B*C*. Likewise, it follows that the angles of one triangle are the supplements of the sides of the other, and that the sides of one are the supplements of the angles of the other.

Hence the perimeter $2p^*$ of A*B*C* is equal to $360° - 2E$, and its spherical excess $2E^*$ is equal to $360° - 2p$, where $2E$ and $2p$ refer to the triangle ABC. In other words half the spherical excess of one triangle is the supplement of half the perimeter of the other.

Both the *triangle of elements* and the *derived triangle* of the polar triangle A*B*C* are thus easily established (Figs. 13

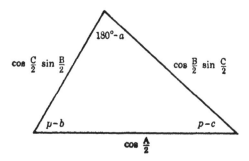

Fɪɢ. 13.—Triangle of elements of the polar triangle (relative to a).

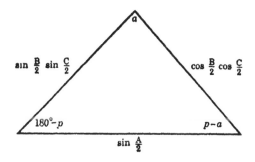

Fɪɢ. 14.—Derived triangle of the polar triangle (relative to a).

and 14). Their sides and angles are functions of the parts of the triangle ABC, in particular of its semi-perimeter p.

Analogous key-triangles, relative to b and c, can be obtained by cyclic permutations.

Remark.—It is easy to remember how to transform the key-triangles of the primitive triangle into those of the polar triangle. Instead of a function of a half-side, write the co-function of the half-angle; thus $\sin \frac{a}{2}$ becomes $\cos \frac{A}{2}$, $\sin \frac{b}{2} \cos \frac{c}{2}$ becomes $\cos \frac{B}{2} \sin \frac{C}{2}$, etc. Instead of an angle, write the supplement of the side; thus A becomes $(180° - a)$, and $(180° - A)$ becomes a. Instead of half the spherical excess, E, write the supplement of half the perimeter $(180° - p)$. Instead of an angle minus half the spherical excess, write half the perimeter minus the side; thus $(A - E)$ becomes $(p - a)$ etc. The latter transformation is immediately apparent, since $A^* - E^* = (180° - a) - (180° - p)$.

4. Derivation of the Formulae of Spherical Trigonometry.

All the formulae of spherical trigonometry are derived from the key-triangles (which have been obtained without any restrictive hypothesis on the spherical triangle, and are therefore perfectly general) by applying to them the known formulae of plane trigonometry. *Each formula of spherical trigonometry can thus be derived independently of the others.*

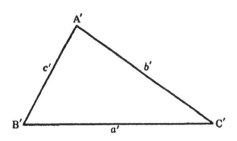

FIG. 15.—A plane triangle.
$$A' + B' + C' = 180°$$
$$a' + b' + c' = 2p'$$

The parts of a plane triangle (Fig. 15) will be designated by primed letters: A′, B′, C′, the angles; a', b', c', the opposite sides; $2p' = a' + b' + c'$, the perimeter.

The parts of a spherical triangle (Fig. 16) will be designated by unprimed letters: A, B, C, the angles; a, b, c, the

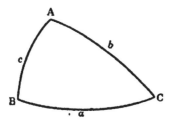

Fig. 16.—A spherical triangle.
$$A + B + C - 180° = 2E$$
$$a + b + c = 2p$$

sides; $2p = a + b + c$, the perimeter; $2E = A + B + C - 180°$, the spherical excess.

EXERCISES

1. Derive the triangle of elements relative to the angle B from that relative to the angle A (Fig. 9) by cyclic permutations.

2. Same question for the triangle of elements relative to C.

3. Draw the derived triangles relative to the angles B and C.

4. Derive, for the polar triangle, the triangle of elements and the derived triangle: (i) relative to the side b, (ii) relative to the side c.

5. Show that the area of the triangle of elements and that of the derived triangle are both equal to $\Delta/8$, where

$$\Delta = 2\sqrt{\sin p \sin(p - a) \sin(p - b) \sin(p - c)}.$$

(Δ, called the *sine of the trihedral angle*, is a useful function of the face angles of the trihedron, which are the sides of the spherical triangle.)

CHAPTER III

HOW THE KEY-TRIANGLES ARE PUT TO WORK

1. Napier's Analogies.[1] Napier's analogies are relations between five parts of a spherical triangle.

$$\frac{\tan \frac{1}{2}(B-C)}{\cot \frac{1}{2}A} = \frac{\sin \frac{1}{2}(b-c)}{\sin \frac{1}{2}(b+c)}, \qquad \frac{\tan \frac{1}{2}(B+C)}{\cot \frac{1}{2}A} = \frac{\cos \frac{1}{2}(b-c)}{\cos \frac{1}{2}(b+c)}; \qquad (1)$$

$$\frac{\tan \frac{1}{2}(b-c)}{\tan \frac{1}{2}a} = \frac{\sin \frac{1}{2}(B-C)}{\sin \frac{1}{2}(B+C)}, \qquad \frac{\tan \frac{1}{2}(b+c)}{\tan \frac{1}{2}a} = \frac{\cos \frac{1}{2}(B-C)}{\cos \frac{1}{2}(B+C)}. \qquad (2)$$

These formulae are read directly from the key-triangles, by applying the Law of Tangents: In a plane triangle, the tangent of the half-difference of two angles is to the tangent of their half-sum (or to the cotangent of half the third angle) as the difference of the opposite sides is to their sum.

The triangle of elements (Fig. 9) yields the first analogy. Note that the half-sum of two angles is B -- C. We have

$$\frac{\tan \frac{B-C}{2}}{\cot \frac{A}{2}} = \frac{\sin \frac{b}{2} \cos \frac{c}{2} - \sin \frac{c}{2} \cos \frac{b}{2}}{\sin \frac{b}{2} \cos \frac{c}{2} + \sin \frac{c}{2} \cos \frac{b}{2}} = \frac{\sin \frac{b-c}{2}}{\sin \frac{b+c}{2}}.$$

The derived triangle (Fig. 11) gives the second analogy. The last two are obtained from the key-triangles (Figs. 13 and 14) of the polar triangle.

2. Delambre's Formulae.[2] Delambre's formulae are relations between all six parts of a spherical triangle.

[1] *Analogies* is an archaic term for *proportions*.
[2] Improperly attributed to Gauss by certain authors.

$$\frac{\cos \tfrac{1}{2}(B-C)}{\sin \tfrac{1}{2}A} = \frac{\sin \tfrac{1}{2}(b+c)}{\sin \tfrac{1}{2}a}, \qquad \frac{\sin \tfrac{1}{2}(B-C)}{\cos \tfrac{1}{2}A} = \frac{\sin \tfrac{1}{2}(b-c)}{\sin \tfrac{1}{2}a}; \qquad (3)$$

$$\frac{\cos \tfrac{1}{2}(B+C)}{\sin \tfrac{1}{2}A} = \frac{\cos \tfrac{1}{2}(b+c)}{\cos \tfrac{1}{2}a}, \qquad \frac{\sin \tfrac{1}{2}(B+C)}{\cos \tfrac{1}{2}A} = \frac{\cos \tfrac{1}{2}(b-c)}{\cos \tfrac{1}{2}a}. \qquad (4)$$

The Law of Sines of plane trigonometry, applied to the triangle of elements (Fig. 9), gives

$$\frac{\sin \dfrac{a}{2}}{\sin A} = \frac{\sin \dfrac{b}{2} \cos \dfrac{c}{2}}{\sin (B - E)} = \frac{\sin \dfrac{c}{2} \cos \dfrac{b}{2}}{\sin (C - E)},$$

which, by the theory of proportions, becomes

$$\frac{\sin \dfrac{a}{2}}{\sin A} = \frac{\sin \dfrac{b+c}{2}}{\sin(B-E) + \sin(C-E)} = \frac{\sin \dfrac{b-c}{2}}{\sin(B-E) - \sin(C-E)}$$

or

$$\frac{\sin \dfrac{a}{2}}{2 \sin \dfrac{A}{2} \cos \dfrac{A}{2}} = \frac{\sin \dfrac{b+c}{2}}{2 \cos \dfrac{A}{2} \cos \dfrac{B-C}{2}} = \frac{\sin \dfrac{b-c}{2}}{2 \sin \dfrac{A}{2} \sin \dfrac{B-C}{2}},$$

and Delambre's first two formulae follow immediately.

The same method, applied to the derived triangle (Fig. 11), yields the last two formulae.

3. Euler's Formula.

Euler's formula expresses E, one-half of the spherical excess,[3] in terms of the sides.

$$\cos E = \frac{1 + \cos a + \cos b + \cos c}{4 \cos \dfrac{a}{2} \cos \dfrac{b}{2} \cos \dfrac{c}{2}}. \qquad (5a)$$

[3] The value E of one-half the spherical excess is useful in calculating the area of a triangle. It is known from geometry (see Appendix 1) that the area of a spherical triangle is to the area of the sphere as E (expressed in degrees) is to 360.

This formula is obtained from the derived triangle (Fig. 11), by applying the Law of Cosines of plane trigonometry

$$a'^2 = b'^2 + c'^2 - 2b'c' \cos A'.$$

We have

$$\sin^2 \frac{b}{2} \sin^2 \frac{c}{2} = \cos^2 \frac{a}{2} + \cos^2 \frac{b}{2} \cos^2 \frac{c}{2} - 2 \cos \frac{a}{2} \cos \frac{b}{2} \cos \frac{c}{2} \cos E,$$

$$2 \cos \frac{a}{2} \cos \frac{b}{2} \cos \frac{c}{2} \cos E = \cos^2 \frac{a}{2}$$
$$+ \left(\cos \frac{b}{2} \cos \frac{c}{2} + \sin \frac{b}{2} \sin \frac{c}{2} \right) \left(\cos \frac{b}{2} \cos \frac{c}{2} - \sin \frac{b}{2} \sin \frac{c}{2} \right),$$

$$4 \cos \frac{a}{2} \cos \frac{b}{2} \cos \frac{c}{2} \cos E = 2 \cos^2 \frac{a}{2} + 2 \cos \frac{b-c}{2} \cos \frac{b+c}{2},$$

from which Euler's formula follows immediately.

4. Lhuilier's Formula. Lhuilier's formula gives the fourth of the spherical excess in terms of the sides (and the semiperimeter p).

$$\tan^2 \frac{E}{2} = \tan \frac{p}{2} \tan \frac{p-a}{2} \tan \frac{p-b}{2} \tan \frac{p-c}{2}. \quad (5b)$$

It can be obtained directly from the derived triangle (Fig. 11) by expressing the tangent of half the angle E in terms of the sides. The plane trigonometry formula is

$$\tan^2 \frac{A'}{2} = \frac{(p'-b')(p'-c')}{p'(p'-a')}.$$

We have

$$2p' - 2b' = \cos \frac{b-c}{2} - \cos \frac{a}{2},$$

$$2p' = \cos \frac{b-c}{2} + \cos \frac{a}{2},$$

and

$$2p' - 2c' = \cos \frac{a}{2} - \cos \frac{b+c}{2} \, .$$

$$2p' - 2a' = \cos \frac{a}{2} + \cos \frac{b+c}{2} \, ,$$

Since

$$\frac{\cos P - \cos Q}{\cos P + \cos Q} = \tan \frac{P+Q}{2} \tan \frac{Q-P}{2} \, ,$$

we are led to

$$\tan^2 \frac{E}{2} = \tan \frac{a+b-c}{4} \tan \frac{a-b+c}{4}$$
$$\times \tan \frac{a+b+c}{4} \tan \frac{b+c-a}{4} \, ,$$

which is Lhuilier's formula, as $2p = a + b + c$.

*5. Cagnoli's Formula. Cagnoli's formula gives the half of the spherical excess in terms of the sides (and the semi-perimeter p).

$$\sin E = \frac{\sqrt{\sin p \, \sin(p-a) \, \sin(p-b) \, \sin(p-c)}}{2 \cos \frac{a}{2} \cos \frac{b}{2} \cos \frac{c}{2}} \, . \qquad (5c)$$

The area T' of the derived triangle (Fig. 11) may be expressed as the half-product of the base by the altitude,

$$T' = \tfrac{1}{2} \cos \frac{a}{2} \cos \frac{b}{2} \cos \frac{c}{2} \sin E,$$

or, in terms of its sides and semi-perimeter,[4]

$$T' = \sqrt{p'(p'-a')(p'-b')(p'-c')}$$
$$= \tfrac{1}{4} \sqrt{\sin p \, \sin(p-a) \, \sin(p-b) \, \sin(p-c)}.$$

[4] Cp. Exercise 5, Chapter II and Section 4, this chapter.

Equating the two expressions immediately gives the desired formula.

*6. The Spherical Excess in Terms of Two Sides and Their Included Angle.   In a plane triangle, as a consequence of the Law of Sines,

$$\cot B' = \frac{c' - b' \cos A'}{b' \sin A'}.$$

This formula, applied to the derived triangle (Fig. 11), —in which the angles $(180° - A)$, E, $(A - E)$ are taken as A', B', C', respectively,—yields the desired relation

$$\cot E = \frac{\cos \dfrac{b}{2} \cos \dfrac{c}{2} + \sin \dfrac{b}{2} \sin \dfrac{c}{2} \cos A}{\sin \dfrac{b}{2} \sin \dfrac{c}{2} \sin A}$$

or

$$\boxed{\cot E = \frac{\cot \dfrac{b}{2} \cot \dfrac{c}{2} + \cos A}{\sin A}.} \tag{6}$$

EXERCISES

1. Express $\cos p$ in terms of the angles, by applying Euler's formula to the polar triangle.

2. Derive the expression of $\cos p$ in terms of the angles from the derived triangle of the polar triangle (Fig. 14), in the same way as Euler's formula has been derived (Sn. 3).

3. From Napier's second analogy (1), derive a formula to calculate E in terms of two sides (b, c) and their included angle (A).

4. Show that, in a right-angled triangle $(C = 90°)$,

$$\tan E = \tan \frac{a}{2} \tan \frac{b}{2}.$$

Hint: use the formula derived in the preceding exercise.

5. Find $\sin \dfrac{E}{2}$ and $\cos \dfrac{E}{2}$ in terms of the sides.   Apply the same method as that used in Sn. 4 for deriving Lhuilier's formula (6).

6. Derive Lhuilier's formula (6) from the results obtained in the preceding exercise.

7. Derive Cagnoli's formula from Lhuilier's formula.

8. Gua's formula gives cot E in terms of the sides. Find what it is.

9. You now know six formulae by means of which, given the sides, the spherical excess can be calculated. Which one would you prefer if you had to depend on logarithmic calculations? Which one would be easiest to use if a calculating machine were available?

10. The sides of a spherical triangle are $a = 35°26'$, $b = 42°15'$, $c = 18°22'$. Calculate the spherical excess by two different formulae.

CHAPTER IV

RELATIONS BETWEEN FOUR PARTS
OF A SPHERICAL TRIANGLE

1. Relation between the Three Sides and One Angle.[1] Applying the Law of Cosines of plane trigonometry

$$a'^2 = b'^2 + c'^2 - 2b'c' \cos A'$$

to the triangle of elements (Fig. 9), we get

$$\sin^2 \frac{a}{2} = \sin^2 \frac{b}{2} \cos^2 \frac{c}{2} + \sin^2 \frac{c}{2} \cos^2 \frac{b}{2}$$
$$- 2 \sin \frac{b}{2} \cos \frac{c}{2} \sin \frac{c}{2} \cos \frac{b}{2} \cos A.$$

In view of

$$2 \sin^2 \frac{x}{2} = 1 - \cos x, \qquad 2 \cos^2 \frac{x}{2} = 1 + \cos x;$$

the above formula, multiplied by 4, may be written

$$2(1 - \cos a) = (1 - \cos b)(1 + \cos c)$$
$$+ (1 - \cos c)(1 + \cos b) - 2 \sin b \sin c \cos A,$$

whence the desired formula

$$\boxed{\cos a = \cos b \cos c + \sin b \sin c \cos A.} \tag{7}$$

This is expressed: The cosine of the side opposite the given angle is equal to the product of the cosines of the other two sides, plus the product of the sines of these two sides times the cosine of the given angle.

[1] Used in Ex. 16, 21, 24, 26, 28 (Ch. VIII).

2. Expression of the Half-Angles in Terms of the Three Sides.[2] The following formulae of plane trigonometry

$$\sin^2 \frac{A'}{2} = \frac{(p' - b')(p' - c')}{b'c'}, \qquad \cos^2 \frac{A'}{2} = \frac{p'(p' - a')}{b'c'},$$

$$\tan^2 \frac{A'}{2} = \frac{(p' - b')(p' - c')}{p'(p' - a')},$$

are applied to the triangle of elements (Fig. 9).

We have

$$2p' = \sin \frac{b + c}{2} + \sin \frac{a}{2} = 2 \sin \frac{p}{2} \cos \frac{p - a}{2},$$

$$2p' - 2a' = \sin \frac{b + c}{2} - \sin \frac{a}{2} = 2 \sin \frac{p - a}{2} \cos \frac{p}{2}, \text{ etc.}$$

Hence, in the spherical triangle,

$$\sin^2 \frac{A}{2} = \frac{\sin(p - b) \sin(p - c)}{\sin b \sin c}, \tag{8a}$$

$$\cos^2 \frac{A}{2} = \frac{\sin p \sin(p - a)}{\sin b \sin c}, \tag{8b}$$

$$\tan^2 \frac{A}{2} = \frac{\sin(p - b) \sin(p - c)}{\sin p \sin(p - a)}. \tag{8c}$$

These formulae are easily remembered on account of their similarity with the corresponding formulae of plane trigonometry.

3. Relation between Two Sides and Their Opposite Angles.[3] Apply the Law of Sines of plane trigonometry to the derived triangle relative to A (Fig. 11).

$$\frac{\cos \frac{a}{2}}{\sin A} = \frac{\sin \frac{b}{2} \sin \frac{c}{2}}{\sin E}.$$

[2] Used in Ex. 1, 18, 19, 20 (Ch. VIII).
[3] Used in Ex. 18, 21, 25, 27 (Ch. VIII).

Multiply both members by $2 \sin \dfrac{a}{2}$,

$$\frac{\sin a}{\sin A} = \frac{2 \sin \dfrac{a}{2} \sin \dfrac{b}{2} \sin \dfrac{c}{2}}{\sin E}.$$

In like manner, from the derived triangle relative to B, we get

$$\frac{\sin b}{\sin B} = \frac{2 \sin \dfrac{a}{2} \sin \dfrac{b}{2} \sin \dfrac{c}{2}}{\sin E},$$

which was to be expected from the symmetrical form of the right-hand member of the above equations. Hence

$$\boxed{\frac{\sin a}{\sin A} = \frac{\sin b}{\sin B}.} \tag{9}$$

This is expressed: The sines of the sides are proportional to the sines of the opposite angles.

4. Relation between the Three Angles and One Side.[4] The Law of Cosines of plane trigonometry is applied to the triangle of elements of the polar triangle (Fig. 13). The method is the same as for the first formula (Sn. 1).

$$\boxed{\cos A = -\cos B \cos C + \sin B \sin C \cos a.} \tag{10}$$

This formula is easily remembered on account of its similarity with formula (7). Note the minus sign in the second member, however.

5. Expression of the Half-Sides in Terms of the Three Angles.[5] The method used in Sn. 2 could be applied to the triangle of elements of the polar triangle (Fig. 13). Another

[4] Used in Ex. 27 (Ch. VIII).
[5] Used in Ex. 19 (Ch. VIII).

method does not require the use of the polar triangle. Draw the two key-triangles relative to B and those relative to C.

The Law of Sines of plane trigonometry, applied to the triangle of elements relative to C, gives

$$\frac{\sin(A - E)}{\sin C} = \frac{\sin \dfrac{a}{2} \cos \dfrac{b}{2}}{\sin \dfrac{c}{2}} \, ;$$

the same law, applied to the derived triangle relative to B, gives

$$\frac{\sin E}{\sin B} = \frac{\sin \dfrac{c}{2} \sin \dfrac{a}{2}}{\cos \dfrac{b}{2}} \, .$$

Multiplying these two relations by each other yields immediately

$$\sin^2 \frac{a}{2} = \frac{\sin E \sin(A - E)}{\sin B \sin C} \, . \tag{11a}$$

Likewise, the Law of Sines, applied to the triangles of elements relative to B and to C, yield two relations: one between B, C − E, and the opposite sides; the other, between C, B − E, and the opposite sides. These two relations, multiplied by each other, give

$$\cos^2 \frac{a}{2} = \frac{\sin(B - E) \sin(C - E)}{\sin B \sin C} \, . \tag{11b}$$

Finally, the Law of Sines may be applied to both key-triangles relative to B, giving the relations

$$\frac{\sin E}{\sin(B - E)} = \frac{\sin \dfrac{c}{2} \sin \dfrac{a}{2}}{\cos \dfrac{c}{2} \cos \dfrac{a}{2}} \, , \qquad \frac{\sin(A - E)}{\sin(C - E)} = \frac{\sin \dfrac{a}{2} \cos \dfrac{c}{2}}{\sin \dfrac{c}{2} \cos \dfrac{a}{2}} \, ,$$

which, multiplied by each other, lead to

$$\tan^2 \frac{a}{2} = \frac{\sin E \sin(A - E)}{\sin(B - E) \sin(C - E)} \cdot \qquad (11c)$$

The last formula could, of course, be derived from the preceding two.

The formulae (11) can only be remembered after careful comparison with the formulae (8). Note the deceiving similarity between (8a) and (11b), etc.

6. Relation between Two Sides, their Included Angle, and the Angle Opposite One of them (that is to say, between four consecutive parts).[6]   Napier's first two analogies, as read from the key-triangles relative to C, give

$$\tan \frac{A - B}{2} = \frac{\sin \dfrac{a - b}{2}}{\sin \dfrac{a + b}{2}} \cot \frac{C}{2},$$

$$\tan \frac{A + B}{2} = \frac{\cos \dfrac{a - b}{2}}{\cos \dfrac{a + b}{2}} \cot \frac{C}{2}.$$

A relation between a, b, C, A can be obtained from these two equations by eliminating B, which is easily done as follows.
Since

$$A = \frac{A - B}{2} + \frac{A + B}{2},$$

we may write

$$\tan A = \frac{\tan \dfrac{A - B}{2} + \tan \dfrac{A + B}{2}}{1 - \tan \dfrac{A - B}{2} \tan \dfrac{A + B}{2}} = \frac{N}{D},$$

[6] Used in Ex. 2, 21, 27 (Ch. VIII).

where

$$N = \left(\frac{\sin \dfrac{a-b}{2}}{\sin \dfrac{a+b}{2}} + \frac{\cos \dfrac{a-b}{2}}{\cos \dfrac{a+b}{2}} \right) \cot \frac{C}{2}$$

$$= \frac{2 \sin a \cos \dfrac{C}{2}}{\sin(a+b) \sin \dfrac{C}{2}}$$

$$= \frac{\sin a \sin C}{\sin(a+b) \sin^2 \dfrac{C}{2}}$$

and

$$D = 1 - \frac{\sin(a-b)}{\sin(a+b)} \cot^2 \frac{C}{2}$$

$$= \frac{\sin(a+b) \sin^2 \dfrac{C}{2} - \sin(a-b) \cos^2 \dfrac{C}{2}}{\sin(a+b) \sin^2 \dfrac{C}{2}} .$$

Substituting for N and D gives

$$\tan A$$

$$= \frac{\sin a \sin C}{\sin a \cos b \left(\sin^2 \dfrac{C}{2} - \cos^2 \dfrac{C}{2} \right) + \sin b \cos a \left(\sin^2 \dfrac{C}{2} + \cos^2 \dfrac{C}{2} \right)},$$

whence

$$\sin b \cos a - \sin a \cos b \cos C = \sin a \sin C \cot A,$$

or, on dividing by $\sin a$ and transposing,

$$\boxed{\cot a \sin b = \cos b \cos C + \sin C \cot A.} \tag{12}$$

This is known as the "cot-sin-cos" formula. (Note the symmetry in the sequence of the trigonometric functions: cot-sin-cos ... cos-sin-cot.) It is expressed: The cotangent of the side opposite one of the given angles times the sine of the other side is equal to the cosine of the latter times the cosine of the included angle, plus the sine of the latter times the cotangent of the angle opposite the first side.

Remark.—Formula (12) is sometimes written in another, just as elegant, form:

$$\cos b \cos C = \sin b \cot a - \sin C \cot A. \qquad (12a)$$

It is then expressed: The product of the cosines of the side and the angle that are *not* opposite any given part is equal to the difference of their sines, each multiplied by a cotangent, respectively that of the given side and the given angle that are opposite each other.

EXERCISES

1. Express $\cos p$ in terms of the angles, by using Delambre's formulae and formula (7). *Hint:*

$$p = \tfrac{1}{2}a + \tfrac{1}{2}(b + c).$$

2. Transform formula (11b) into formula (10). *Hint:* To eliminate E multiply both members by 2 and change the product of two sines in the numerator into a difference of two cosines.

3. Derive formula (10) by applying formula (7) to the polar triangle.

4. Derive the three formulae (11) by means of the triangle of elements of the polar triangle.

5. Derive formula (12) from Napier's last two analogies (2).

6. Write the formula giving $\cos c$ in terms of a, b, C. Then replace $\cos c$ in the formula (7) by the value just found and, using the proportionality of the sines of sides and opposite angles, derive formula (12).

7. Let 2S designate the sum of the angles of a spherical triangle. What form will the three formulae (11) assume with this notation?

8. From formulae (8a) and (8b), find an expression of sin A. Show that it is equivalent to

$$\sin A = \frac{1}{\sin b \sin c} \sqrt{1 - \cos^2 a - \cos^2 b - \cos^2 c + 2 \cos a \cos b \cos c}.$$

*9. The expression of sin A, in terms of the sides, found from formulae (8a) and (8b) can also be obtained directly from the triangle of elements (Fig. 9), by applying to it the method used for deriving Cagnoli's formula (Ch. III, Sn. 5).

CHAPTER V

RIGHT-ANGLED TRIANGLES

1. The Right-Angled Triangle, a Special Case of the Oblique-Angled Triangle. The formulae established in the preceding chapter give relations between four parts of a triangle. Of these parts, at least one is necessarily an angle. By letting an angle equal 90°, a formula derived for the general case of an oblique-angled triangle is transformed into a relation between three parts (other than the right angle) of a right-angled triangle. By convention [1] the triangle ABC is made right-angled at C. It may be necessary to rearrange the general formulae by permutation of letters, so that the angle which is to become the right angle be labeled C. The side opposite C will therefore be the hypotenuse c.

There are six different ways of choosing three parts from the five parts (other than the right angle) of a right-angled triangle. The hypotenuse may be taken, either with the other two sides, or with the two angles other than the right angle, or with one side and one angle. In the latter case, the angle may be opposite the chosen side or adjacent to it. If the hypotenuse is not chosen, there are only two possibilities: two angles and one side, necessarily opposite one of the angles; or two sides and one angle, necessarily opposite one of the sides.

Right-angled triangles are always with us. The six relations about to be derived have therefore proved to be very useful in practice. You will find it advantageous to commit them to memory, preferably in the form of statements rather than equations.

---

[1] This convention is not universal. Many authors make A = 90°.

2. Relations between the Hypotenuse and Two Sides or Two Angles.[2] *The Product Formulae.*

Formula (7) may be written

$$\cos c = \cos a \cos b + \sin a \sin b \cos C.$$

It becomes,[3] for C = 90°,

$$\boxed{\cos c = \cos a \cos b.} \tag{13}$$

Formula (10) may be written

$$\cos C = -\cos A \cos B + \sin A \sin B \cos c.$$

It becomes [4]

$$\boxed{\cos c = \cot A \cot B.} \tag{14}$$

This is expressed: The **cosine of the hypotenuse** is equal to the product of the cosines of the two sides, or the product of the cotangents of the two angles.

3. Relations between the Hypotenuse, One Side, and One Angle. *The Ratios of Sines and Tangents.*

Formula (9) may be written

$$\frac{\sin c}{\sin C} = \frac{\sin a}{\sin A}.$$

It becomes,[5] for C = 90°,

$$\sin a = \sin c \sin A$$

---

[2] The terms *sides* and *angles*, when applied to a right-angled triangle, are construed to mean *sides other than the hypotenuse* and *angles other than the right angle.*
[3] Formula (13) is used in Ex. 3, 7, 15, 17, 29 (Ch. VIII).
[4] Formula (14) is used in Ex. 9, 12, 26 (Ch. VIII).
[5] Formula (15) is used in Ex. 4, 14, 23, 24, 29 (Ch. VIII).

or

$$\boxed{\frac{\sin a}{\sin c} = \sin A.} \tag{15}$$

Formula (12) may be written

$$\cot c \sin a = \cos a \cos B + \sin B \cot C.$$

It becomes [6]

$$\cos B = \cot c \tan a$$

or

$$\boxed{\frac{\tan a}{\tan c} = \cos B.} \tag{16}$$

With respect to the side a, note that A is the *opposite angle*, whereas B is the *adjacent angle*.

The relations are remembered as follows. In each case consider the side and the hypotenuse; the ratio of their sines is equal to the sine of the opposite angle, the ratio of their tangents is equal to the *cosine* of the adjacent angle.

Remark.—Compare the definitions of sine and cosine in a plane right-angled triangle: $\sin A' = \cos B' = a'/c'$.

4. Relations between Two Angles and One Side, or between Two Sides and One Angle. *The Ratios of Cosines and Cotangents.*

Formula (10),

$$\cos A = - \cos B \cos C + \sin B \sin C \cos a,$$

becomes,[7] on letting $C = 90°$,

$$\cos A = \sin B \cos a$$

or

$$\boxed{\frac{\cos A}{\cos a} = \sin B.} \tag{17}$$

---

[6] Formula (16) is used in Ex. 4, 8, 20 (Ch. VIII).
[7] Formula (17) is used in Ex. 13 (Ch. VIII).

Formula (12),

$$\cot a \sin b = \cos b \cos C + \sin C \cot A,$$

becomes,[8] for C = 90°,

$$\cot a \sin b = \cot A$$

or

$$\boxed{\frac{\cot A}{\cot a} = \sin b.}$$ (18)

The relations are remembered as follows. In each case consider an angle and its opposite side; the ratio of their **cosines** is equal to the sine of the other **angle**, the ratio of their **cotangents** is equal to the sine of the other **side**.

Remark.—Notice that all but one of the four ratios considered in the last two sections give the *sine* of a part.

5. Napier's Rule. Although some people prefer to memorize the preceding formulae as such, Napier's rule, which includes all the possible relations between any three parts of a right-angled triangle, may be found useful by others.

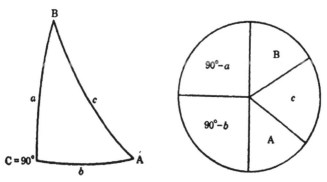

Fig. 17.—Napier's rule.

In the triangle ABC, right-angled at C, ignore the right angle; then, the hypotenuse, the other two angles, and the

[8] Formula (18) is used in Ex. 7, 15, 17 (Ch. VIII).

complements of the sides about the right angle are Napier's *five cyclic parts.* The rule is expressed as follows.

The **cosine** of any of the five parts is equal to the product of the **sines** of the opposite parts or the product of the **cotangents** of the adjacent parts.

$$\cos c = \cos a \cos b = \cot A \cot B,$$
$$\cos A = \sin B \cos a = \cot c \tan b,$$
$$\cos B = \sin A \cos b = \cot c \tan a,$$
$$\sin a = \sin c \sin A = \cot B \tan b,$$
$$\sin b = \sin c \sin B = \cot A \tan a.$$

This amazing rule is more than just a mnemonic trick. It is a theorem, which was given with a separate proof [9] by Napier.

It is easy to check that the ten relations yielded by the rule correspond, with some duplication, to the formulae (13) to (18).

[9] Napier's proof is outside the scope of this book.

CHAPTER VI

EXAMPLES OF CALCULATIONS

1. Solution of Triangles.[1]

(i) *Oblique-angled triangles.*

Three parts being given, any fourth part may be calculated by one of the formulae (7) to (12). To find an angle in terms of the three sides, use one of the formulae (8), giving the half-angle. Likewise, to find a side in terms of the three angles, use one of the formulae (11), giving the half-side. In all other cases, use one of the fundamental formulae (7), (9), (10), (12).

(ii) *Right-angled triangles.*

Two parts being given, besides the right angle, any third part may be calculated by one of the formulae (13) to (18). Find the appropriate formula or use Napier's rule.

An isosceles triangle is divided into two right-angled triangles by an arc drawn from the vertex at right angles to the base; this arc bisects the base and the opposite angle. Isosceles triangles may also be solved like oblique-angled triangles by means of the fundamental formulae; formulae (8) and (11) should be avoided, as they would complicate the calculations.

(iii) *Right-sided (or quadrantal) triangles.*

The polar triangle of a right-sided triangle is right-angled; Napier's rule may be used to solve the polar triangle, from which the parts of the quadrantal triangle are then computed.

---

[1] We have seen (Introduction, Sn. 2) that any angle or any side of a spherical triangle is less than 180°. It follows that a case of ambiguity will present itself whenever a part is to be calculated by a formula that gives its **sine** (two supplementary angles having equal sines).

Right-sided triangles are also easily solved by the fundamental formulae, which are greatly simplified when one of the sides is equal to 90°.

2. Formulae Adapted to Logarithmic Computation.

(i) FORMULA: $\cos a = \cos b \cos c + \sin b \sin c \cos A$.

The unknown may be either a, b (or c), or A. If the unknown is a, factorize $\cos b$:

$$\cos a = \cos b(\cos c + \tan b \sin c \cos A).$$

Introduce an auxiliary angle u so as to arrive at the sine of the sum of two angles.

Let
$$\cot u = \tan b \cos A,$$
then
$$\cos a = \frac{\cos b}{\sin u}(\sin u \cos c + \sin c \cos u)$$
or
$$\cos a = \frac{\cos b \sin(c + u)}{\sin u}.$$

If the unknown is c, the same method gives immediately

$$\sin(c + u) = \frac{\cos a \sin u}{\cos b}.$$

If the unknown is A, use the formulae (8), which are adapted to logarithmic computation.

(ii) FORMULA: $\cot a \sin b = \cos b \cos C + \sin C \cot A$.

If the unknown is a, we write

$$\cot a = \cot b\left(\cos C + \frac{\sin C \cot A}{\cos b}\right).$$

Let
$$\cot v = \frac{\cot A}{\cos b},$$

then

$$\cot a = \frac{\cot b}{\sin v} (\sin v \cos C + \sin C \cos v)$$

or

$$\cot a = \frac{\cot b \sin(C + v)}{\sin v}.$$

If the unknown is C, the same method gives immediately

$$\sin(C + v) = \frac{\cot a \sin v}{\cot b}.$$

If the unknown is b, we write

$$\cot a \left(\sin b - \cos b \frac{\cos C}{\cot a} \right) = \sin C \cot A.$$

Let

$$\tan w = \frac{\cos C}{\cot a},$$

then

$$\frac{\cot a}{\cos w} (\sin b \cos w - \sin w \cos b) = \sin C \cot A$$

or

$$\sin(b - w) = \frac{\sin C \cot A \cos w}{\cot a}.$$

If the unknown is A, the same method gives immediately

$$\cot A = \frac{\cot a \sin(b - w)}{\sin C \cos w}.$$

(iii) FORMULA: $\cos A = - \cos B \cos C + \sin B \sin C \cos a$.

The method is the same as for the first formula (Sn. 2, i).
The results follow.

Let

$$\cot x = \tan B \cos a.$$

If the unknown is A,

$$\cos A = \frac{\cos B \sin(C - x)}{\sin x}.$$

If the unknown is C,

$$\sin(C - x) = \frac{\cos A \sin x}{\cos B}.$$

If the unknown is a, use the formulae (11), which are adapted to logarithmic computation.

3. Numerical Calculations.

Computations are carried out either by means of logarithms or with a calculating machine (in which case tables of natural values of the trigonometric functions are necessary). In either case, a neat calculation form is essential.

(i) *By logarithms.*

First write across the top of the page the formula to be employed (adapted to logarithmic computation, if desired). In this formula always designate the parts of the triangle by letters (*not* by their actual values).

Then write the given parts near the left margin of the sheet. Avoid writing the word *log* as much as possible, use it only in front of those logarithms which may be needed again in the course of the calculations. Write the *minus* sign over a negative characteristic,[2] so as to avoid the cum-

---

[2] Calculations with negative figures are quite straightforward, once you become used to them. See for yourself:

```
      1.67                    2̄.67
   2)3.34                  2)3̄.34
      2                       4̄
     ──                      ──
      13                      13
      12                      12
     ──                      ──
      14                      14
      14                      14
     ──                      ──
```

In the division on the left, you say: 2 in 3 goes 1, $1 \times 2 = 2$, $3 - 2 = 1$, bring down 3, 2 in 13 $\cdots$ etc. In the division on the right, you say: 2 in 3 goes 2̄, $2̄ \times 2 = 4̄$, $3 - 4̄ = 1$, bring down 3, 2 in 13 $\cdots$ etc. The only difference is that, in the second case, you take the first partial quotient *by excess* in order to get a positive remainder.

bersome addition and subtraction of 10. In most cases there is no advantage in adding the cologarithm rather than subtracting the logarithm itself; in a simple division, especially, this amounts to making a subtraction and an addition instead of one subtraction only.

Write the result near the left margin of the page.

A second formula may be used as a check. This can be done in two ways: either the solution is carried out in duplicate, by means of two formulae; or a relation between the given parts and the result is verified afterwards. In the course of long calculations (involving a whole chain of triangles, for instance) it is well to check the results from time to time, before proceeding.

(ii) *With the calculating machine.*

Write the formula across the page. Copy only those natural values that may be used again in the course of the calculations. Write the data and the result near the left margin of the page.

4. First Example.

Given the three sides of a triangle, solve for one angle. Use a check formula.

Again compare these two additions of logarithms:

$$
\begin{array}{ll}
9.412\ 5062 \quad -10 & \bar{1}.412\ 5062 \\
8.803\ 7253 \quad -10 & \bar{2}.803\ 7253 \\
\hline
18.216\ 2315 \quad -20 & \bar{2}.216\ 2315 \\
\end{array}
$$

The last partial addition, in the example on the right, reads: 1 (carried over) and $\bar{1}$ make 0, and $\bar{2}$ make $\bar{2}$. Is this really difficult?

(A) SOLUTION BY LOGARITHMS

$$\sin^2\frac{A}{2} = \frac{\sin (p-b)\sin (p-c)}{\sin b \sin c}, \qquad \cos^2\frac{A}{2} = \frac{\sin p \sin(p-a)}{\sin b \sin c}.$$

$$
\begin{aligned}
a &= 50°48'20'' \\
b &= 116°44'50'' \quad (63°15'10'') \dots\dots\dots\dots\dots\dots \quad \overline{1}.950\ 8518 \\
c &= 129°11'40'' \quad (50°48'20'') \dots\dots\dots\dots\dots \quad \overline{1}.889\ 3049 \\
\end{aligned}
$$

$$\overline{1}.840\ 1567$$

$$2p = 296°44'50''$$

$$
\begin{aligned}
& && 171 \\
p &= 148°22'25'' \quad (31°37'35'') \dots\dots \quad \overline{1}.719\ 6275 \\
p-a &= 97°34'5'' \quad (82°25'55'') \dots\dots \quad \overline{1}.996\ 1989 \\
& && 14 \\
\end{aligned}
$$

$$\overline{1}.715\ 8449$$
$$-\ \overline{1}.840\ 1567 \leftarrow$$

$$\log \cos \frac{A}{2} = \frac{\overline{1}.875\ 6882}{\overline{1}.937\ 8441}$$

$$
\begin{aligned}
& 8462 \\
\hline
& 21 \\
& 12\ 1 \\
\hline
& 8\ 9
\end{aligned}
$$

$$\frac{A}{2} = 29°55'41.7''$$

$$
\begin{aligned}
p-b &= 31°37'35'' \dots \overline{1}.719\ 6446 \\
p-c &= 19°10'45'' \dots \overline{1}.516\ 5358 \\
& \phantom{= 19°10'45'' \dots \overline{1}.5} 303 \\
\end{aligned}
$$

$$\overline{1}.236\ 2107$$
$$-\ \overline{1}.840\ 1567 \leftarrow$$

$$\log \sin \frac{A}{2} = \frac{\overline{1}.396\ 0540}{\overline{1}.698\ 0270}$$

$$
\begin{aligned}
& 0204 \\
\hline
& 66 \\
& 36\ 6 \\
\hline
& 29\ 4
\end{aligned}
$$

$$\frac{A}{2} = 29°55'41.8''$$

$A = 59°51'23.6''$ The answer to the closest second is $59°51'24''$.

Suggestion —To find the supplement of an angle, write down the figures from left to right as you mentally increase the

given angle to 179°59′(50/10)″.   Example: given 148°22′25″.
Write 31° while saying 179,   37′ while saying 59, 3 while
saying 50,   and 5″ while saying 10.

(B) SOLUTION BY NATURAL VALUES AND CALCULATING MACHINE

$$\cos A = \frac{\cos a - \cos b \cos c}{\sin b \sin c}.$$

	cos	sin
$a = 50°48′20″$	$0.631\ 9542$	
$b = 116°44′50″$	$-0.450\ 0551$	$0.893\ 0008$
$c = 129°11′40″$	$-0.631\ 9542$	$0.775\ 0058$

$$\cos A = 0.502\ 1668$$
$$1817$$
$$\overline{149}$$
$$125\ 7$$

A   59°51′23.6″. $\overline{23}$ 3

Check formula: $\tan^2 \dfrac{A}{2} = \dfrac{\sin(p-b)\sin(p-c)}{\sin p \sin(p-a)}.$

$2p = 296°44′50″$		
$p = 148°22′25″$ (31°37′35″)...................		$0.524\ 3575$
$p - a = \ \ 97°34′\ 5″$ (82°25′55″) ... $0.991\ 2859$ ...		206
32		$\overline{0.524\ 3781}$
$\overline{0.991\ 2891}$		

$p - b = \ \ 31°37′35″$		$0.524\ 3781$
$p - c = \ \ 19°10′45″$ .... $0.328\ 5003$		
229		
$\overline{0.328\ 5232}$		

$$\tan^2 \frac{A}{2} = 0.331\ 4101$$

$$\tan \frac{A}{2} = 0.575\ 6824$$
$$6708$$
$$\overline{116}$$
$$64\ 6$$
$$\frac{A}{2} = 29°55′41.8″ \qquad \overline{51}\ 4$$

A = 59°51′23.6″.

5. Second Example.

Given the three angles of a triangle, solve for the three sides.

(A) SOLUTION BY LOGARITHMS

$$\sin^2\frac{a}{2} = \frac{\sin E \sin(A-E)}{\sin B \sin C}, \qquad \sin^2\frac{b}{2} = \frac{\sin E \sin(B-E)}{\sin C \sin A},$$

$$\sin^2\frac{c}{2} = \frac{\sin E \sin(C-E)}{\sin A \sin B}.$$

A =	66°47′ 0″	... Ī.963 3253	... Ī.963 3253	
B =	42°30′40″	..Ī.829 7752	... Ī.829 7752	
C =	97°20′30″	..Ī.996 4249 ... Ī.996 4249		
206°38′10″	Ī.826 2001	Ī.959 7502	Ī.793 1005	

$$2E = 26°38′10″$$

E =	13°19′ 5″	..Ī.362 3558 ... Ī.362 4003	... Ī.362 4003	
		445		
A − E =	53°27′55″	..Ī.904 9760		
		78		
B − E =	29°11′35″	.... Ī.688 1819		
		188		
C − E =	84° 1′25″	....	.. Ī.997 6331	

$$
\begin{array}{ccc}
\overline{\text{Ī.267 3841}} & \overline{\text{Ī.050 6010}} & \overline{\text{Ī.360 0334}} \\
-\text{Ī.826 2001} & -\text{Ī.959 7502} & -\text{Ī.793 1005} \\
\hline
\overline{\text{Ī.441 1840}} & \overline{\text{Ī.090 8508}} & \overline{\text{Ī.566 9329}} \\
\end{array}
$$

$$\log\sin.\ \frac{a}{2}\ \text{Ī.720 5920} \qquad .\frac{b}{2}\ \text{Ī.545 4254} \qquad .\frac{c}{2}\ \text{Ī.783 4664}$$

$$
\begin{array}{ccc}
5834 & 3938 & 4575 \\
\hline
86 & 316 & 89 \\
68\ \ 2 & 280\ \ 5 & 82\ \ 8 \\
\hline
17\ \ 8 & 35\ \ 5 & 6\ \ 2 \\
\end{array}
$$

$$\frac{a}{2}=31°42′12.5″ \quad \frac{b}{2}=20°33′15.6″ \quad \frac{c}{2}=37°24′3.2″$$

a =	63°24′25.0″
b =	41° 6′31.2″
c =	74°48′ 6.4″
2p =	179°19′ 2.6″

Check formula: $\tan^2\dfrac{E}{2} = \tan\dfrac{p}{2}\tan\dfrac{p-a}{2}\tan\dfrac{p-b}{2}\tan\dfrac{p-c}{2}.$

$$p = 89°39'31.3'' \qquad \tfrac{1}{2}p = 44°49'45.6''....\overline{1}.997\ 3891$$
$$211$$
$$25\ 3$$
$$p - a = 26°15'\ 6.3'' \quad \tfrac{1}{2}(p - a) = 13°\ 7'33.1'' \quad \overline{1}.367\ 6676$$
$$285\ 6$$
$$9\ 5$$
$$p - b = 48°33'\ 0.1'' \quad \tfrac{1}{2}(p - b) = 24°16'30.0'' \quad \overline{1}.654\ 1690$$
$$p - c = 14°51'24.9'' \quad \tfrac{1}{2}(p - c) = 7°25'42.4'' \quad \overline{1}.115\ 1788$$
$$328$$
$$65\ 6$$
$$\overline{2.134\ 4970}$$

$$\frac{E}{2} = 6°39'32.4'' \qquad \log \tan \frac{E}{2} = \overline{1}.067\ 2485$$
$$2041$$
$$\overline{444}$$
$$E = 13°19'\ 4.8'' \qquad\qquad\qquad 366$$
$$\overline{78\ 0}$$

(B) SOLUTION BY NATURAL VALUES AND CALCULATING MACHINE

$$\cos a = \frac{\cos A + \cos B \cos C}{\sin B \sin C}, \qquad \cos b = \frac{\cos B + \cos C \cos A}{\sin C \sin A},$$
$$\cos c = \frac{\cos C + \cos A \cos B}{\sin A \sin B}.$$

		cos	sin
A =	66°47' 0''	0.394 2093	0.919 0207
B =	42°30'40''	0.737 1463	0.675 7332
C =	97°20'30''	−0.127 7859	0.991 8018
	206°38'10''		

$$a = 63°24'25.0'' \qquad 0.447\ 6505$$
$$6724$$
$$\overline{219}$$
$$217$$
$$\overline{2}$$

$$b = 41°\ 6'31.3'' \qquad 0.753\ 4637$$
$$4678$$
$$\overline{41}$$
$$31\ 9$$
$$\overline{9\ 1}$$

$$c = 74°48'\ 6.5'' \qquad 0.262\ 1589$$
$$1892$$
$$\overline{303}$$
$$280\ 8$$
$$\overline{22\ 2}$$

Check formula:

$$\cos E = \frac{1 + \cos a + \cos b + \cos c}{4 \cos \dfrac{a}{2} \cos \dfrac{b}{2} \cos \dfrac{c}{2}}.$$

	0.850 7602
	177 8
	12 7
$\frac{1}{2}a = 31°42'12.5''$	0.850 7792
	0.936 3322
	68
	6 8
$\frac{1}{2}b = 20°33'15.6''$	0.936 3397
	0.794 3852
	176 4
	23 5
$\frac{1}{2}c = 37°24' 3.2''$	0.794 4052

$$\cos E = 0.973\ 1061$$
$$1119$$
$$58$$
$$56$$
$$2$$

$$E = 13°19' 5.2''$$

$$2E = 26°38'10.4''.$$

6. Third Example.

In a triangle ABC, right-angled at C, the following parts are known: $a = 59°36'30''$, $b = 64°22'$. Find the angle A.

(A) SOLUTION BY LOGARITHMS

$$\sin b = \cot A \tan a,$$

hence

$$\tan A = \frac{\tan a}{\sin b}.$$

$a = 59°36'30''$	....	0.231 7312
$b = 64°22'$	....	$-\overline{1}.955\ 0047$
	$\log \tan A =$	0.276 7265
		167
		98

$$A = 62°\ 7'52''.$$

$$\tan A = \frac{\tan a}{\sin b} = \frac{1.705\ 0269}{0.901\ 5810} = 1.891 \begin{array}{r} 1522 \\ 1094 \\ \hline 428 \end{array}$$

$A = 62°\ 7'52''.$

7. Fourth Example.

Given the geographical coordinates of two points A and B on a sphere, find the spherical distance between these two points.

Let L and M, L' and M' be the longitudes and latitudes of the points A and B, respectively (Fig. 18). Let x be the arc AB.

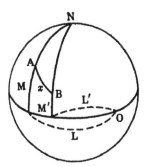

FIG. 18.—Geographical problem.

The formula (7), applied to the triangle NAB, immediately solves the problem.

$$\cos x = \sin M \sin M' + \cos M \cos M' \cos(L - L').$$

If the calculation is to be made by logarithms, write

$$\cos x = \sin M'[\sin M + \cos M \cot M' \cos(L - L')].$$

Let

$$\cot M' \cos(L - L') = \tan w$$

and compute the auxiliary angle w.

We have then

$$\cos x = \frac{\sin M'}{\cos w}(\sin M \cos w + \cos M \sin w)$$

$$= \frac{\sin M' \sin(M + w)}{\cos w}.$$

8. Numerical Application.

Take the following geographical coordinates: New York, lat. 40°43′N., long. 74°0′W.; San Francisco, lat. 37°48′N., long. 122°24½′W. Assume the earth to be spherical and take its radius to be 3960 miles.

What is the spherical distance (in degrees and minutes) between New York and San Francisco? What is the actual distance (in miles)? The coordinates being accurate to half a minute, with what precision can you give the distance?

(A) SOLUTION BY LOGARITHMS

Let L, L′ be the longitudes of San Francisco and New York. Let M, M′ be their latitudes.

$$\tan w = \cot M' \cos(L - L'), \qquad \cos x = \frac{\sin M' \sin(M + w)}{\cos w}.$$

M′	40°43′	....	0.065 1775
L	122°24′30″		
L′	74° 0′ 0″		
L − L′	48°24′30″	....	$\overline{1}$.822 0487
			$\overline{1}$.887 2262
			2024
			$\overline{}$238
			217 5
			$\overline{}$20 5
w	37°38′35.5″		$\overline{1}$.898 6243
M	37°48′		65 2
M + w	75°26′35.5″		$\overline{}$8 2
		log cos w =	$\overline{1}$.898 6316

M′ Ī.814 4600
M + w Ī.985 8270
                                            27 5
                                             2 8
                                    ‾‾‾‾‾‾‾‾‾‾
                                    Ī.800 2900
                              − Ī.898 6316
               log cos x = Ī.901 6584
                                    6649
                                    ‾‾‾‾
                                    65
x = 37° 7′14.1″   63 6
                                    ‾‾‾
                                    1 4

The answer to the nearest half-minute is 37° 7′.

Length of the great circle on the earth (R = 3960 miles).

        2 .... 0.301 0300
        π 0.497 1499
        R 3.597 6952
        log 2πR = ‾‾‾‾‾‾‾‾‾‾‾‾‾
                       4.395 8751
                       8678
                       ‾‾‾‾
                       73
                       70
                       ‾‾
2πR = 24,881.42 miles   3

Length of 1°, 1′, 1″

	1°	1′	1″
2πR	.... 4.395 8751	4.395 8751	4.395 8751
360°	....−2.556 3025		
21,600′	....	−4.334 4538	
1,296,000″	....	....	−6.112 6050
	‾‾‾‾‾‾‾‾‾‾‾	‾‾‾‾‾‾‾‾‾‾‾	‾‾‾‾‾‾‾‾‾‾‾
	1.839 5726	0.061 4213	2̄.283 2701
37°	.... 1.568 2017		
7′	....		0.845 0981	
14.1″	....		1.149 2191
	‾‾‾‾‾‾‾‾‾‾‾	‾‾‾‾‾‾‾‾‾‾‾	‾‾‾‾‾‾‾‾‾‾‾
	3.407 7743	0.906 5194	Ī.432 4892
	7647	82	83
	‾‾‾‾	‾‾‾‾	
	96	12	

37° = 2,557.25   The error on ½′ is more than ½ mile.
7′ = 8.0634
14.1″ = 0.2707   The answer to the nearest half-mile
‾‾‾‾‾‾‾‾‾‾‾   is 2565½ miles.
2,565.58

(B) SOLUTION BY NATURAL VALUES AND CALCULATING MACHINE

$$\cos x = \sin M \sin M' + \cos M \cos M' \cos(L - L').$$

M	37°48′ ..	0.612 9071	0.790 1550	L =	122°24′30″
M′	40°43′ ..	× 0.652 3189	× 0.757 9446	L′ =	74° 0′ 0″
			× 0.663 8174	L − L′ =	48°24′30″

$$\cos x = 0.797\ 3669$$
$$3792$$
$$\overline{123}$$
$$117\ 2$$
$$x = 37°\ 7′14.2″ \qquad \overline{5\ 8}$$

Tables give the lengths of arcs to the radius 1.

37°	.... 0.645 7718		
7′	.... 0.002 0362	½′	0.000 1454
14″	.... 0.000 0727		× 3960
	0.647 8807	½′ =	0.5758 miles
	× 3960		
	x = 2565.6		

The answer to the nearest half-mile is 2565½ miles.

CHAPTER VII

PROBLEMS

1. Let ABC be a tri-rectangular triangle (all angles and sides equal to 90°); let p, q, r be the spherical distances from any point P inside the triangle to the vertices A, B, C, respectively (Fig. 19). Prove that

$$\cos^2 p + \cos^2 q + \cos^2 r = 1.$$

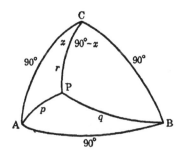

Fɪɢ. 19.—Distances from a point inside a tri-rectangular triangle to the vertices.

In the triangle CAP, by formula (7),

$$\cos p = \sin r \cos x.$$

Likewise, in the triangle BCP,

$$\cos q = \sin r \sin x.$$

Squaring and adding, in order to eliminate x, we get

$$\cos^2 p + \cos^2 q = \sin^2 r = 1 - \cos^2 r,$$

whence the desired relation.

2. A point P is located inside a tri-rectangular triangle ABC. Let x, y, z be its distances from the sides (Fig. 20). Prove that

$$\sin^2 x + \sin^2 y + \sin^2 z = 1.$$

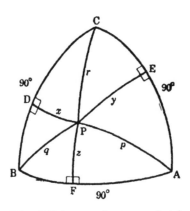

Fig. 20.—Distances from a point inside
a tri-rectangular triangle to the sides.

Join PA, PB, PC and let these arcs be p, q, r.

Let the arcs x, y, z intersect the sides BC, CA, AB in D, E, F. Let BD $= u$, CE $= v$, AF $= w$.

The two right-angled triangles BDP and CDP yield the relations:

$$\cos q = \cos x \cos u, \qquad \cos r = \cos x \sin u.$$

On squaring and adding, we get

$$\cos^2 x = \cos^2 q + \cos^2 r.$$

Likewise,

$$\cos^2 y = \cos^2 r + \cos^2 p,$$
$$\cos^2 z = \cos^2 p + \cos^2 q.$$

On addition, and in view of the fact that $\cos^2 p + \cos^2 q + \cos^2 r = 1$, these three equations give

$$3 - (\sin^2 x + \sin^2 y + \sin^2 z) = 2,$$

whence the desired relation.

3. Let p, q, r and p', q', r' be angles which two straight lines OP and OP′ make with Cartesian coordinate axes (Fig. 21). Prove that the angle x between these two directions is given by

$$\cos x = \cos p \cos p' + \cos q \cos q' + \cos r \cos r'.$$

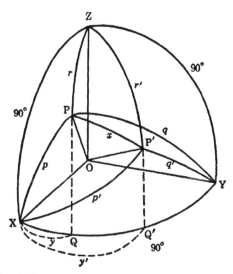

FIG. 21.—Angle between two directions (perspective drawing).

Consider a sphere drawn with the origin O as center. Let the radius of that sphere be taken as the unit of length. Produce the arcs ZP and ZP′ to their intersections, Q and Q′, with XY. Let $XQ = y$, $XQ' = y'$.

In the triangle PZP′,

$$\cos x = \cos r \cos r' + \sin r \sin r' \cos (y' - y),$$

or

$$\cos x = \cos r \cos r' + \sin r \sin r' \cos y' \cos y$$
$$+ \sin r \sin r' \sin y' \sin y.$$

In the triangle XPQ:

$$\cos p = \sin r \cos y.$$

In the triangle XP'Q':

$$\cos p' = \sin r' \cos y'.$$

Likewise, in the triangles YPQ and YP'Q':

$$\cos q = \sin r \sin y, \qquad \cos q' = \sin r' \sin y'.$$

Hence

$$\cos x = \cos r \cos r' + \cos p \cos p' + \cos q \cos q'.$$

4. In a triangle ABC (Fig. 22), an arc x, drawn through one of the vertices C, determines on the opposite side c the arcs BD $= p$ and AD $= q$. Prove that x is given by

$$\sin c \cos x = \sin p \cos b + \sin q \cos a.$$

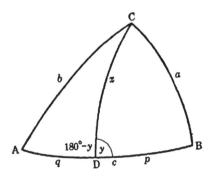

Fig. 22.—Arc through one of the vertices.

Let y designate the angle CDB.   We have

$$\cos a = \cos p \cos x + \sin p \sin x \cos y,$$
$$\cos b = \cos q \cos x - \sin q \sin x \cos y.$$

Multiply the first equation by $\sin q$, the second by $\sin p$, and add the two together, so as to eliminate y.   We get

$$\sin q \cos a + \sin p \cos b = \sin (p + q) \cos x,$$

which is the desired relation; and which may also be written in the form

$$\cos x = \frac{\sin p \cos b + \sin q \cos a}{\sin c}.$$

Applications.

(i) *A median in terms of the sides.*

In this case, $p = q = \frac{1}{2}c$. The general formula becomes

$$\cos x = \frac{\sin \dfrac{c}{2}(\cos a + \cos b)}{\sin c} = \frac{\cos \dfrac{a+b}{2} \cos \dfrac{a-b}{2}}{\cos \dfrac{c}{2}}.$$

(ii) *A bisector in terms of the sides.*

In this case, angle ACD = angle BCD = $\frac{1}{2}C$. We have

$$\frac{\sin p}{\sin \frac{1}{2}C} = \frac{\sin a}{\sin y} \quad \text{and} \quad \frac{\sin q}{\sin \frac{1}{2}C} = \frac{\sin b}{\sin y},$$

whence

$$\frac{\sin p}{\sin q} = \frac{\sin a}{\sin b}.$$

Now, since $p + q = c$,

$$\frac{\sin(c - q)}{\sin q} = \frac{\sin a}{\sin b},$$

which leads to

$$\cot q = \frac{\sin a + \sin b \cos c}{\sin b \sin c}.$$

Hence

$$\sin q = \frac{1}{\sqrt{1 + \cot^2 q}} = \frac{\sin b \sin c}{\sqrt{\sin^2 a + \sin^2 b + 2 \sin a \sin b \cos c}}.$$

Likewise (making q into p and interchanging a and b),

$$\sin p = \frac{\sin a \sin c}{\sqrt{\sin^2 a + \sin^2 b + 2 \sin a \sin b \cos c}}.$$

On substitution, the general formula becomes

$$\cos x = \frac{\sin (a + b)}{\sqrt{\sin^2 a + \sin^2 b + 2 \sin a \sin b \cos c}}.$$

(iii) *An altitude in terms of the sides.*

In this case, $y = 180° - y = 90°$. The general formula, thanks to the formulae of right-angled triangles

$$\cos a = \cos x \cos p \qquad \text{and} \qquad \cos b = \cos x \cos q,$$

can be simplified as follows

$$\cos x = \frac{\cos a}{\cos p} = \frac{\cos b}{\cos q}.$$

From the relation

$$\frac{\cos p}{\cos q} = \frac{\cos a}{\cos b},$$

by transformations similar to those used in (ii), we deduce

$$\tan q = \frac{\cos a - \cos b \cos c}{\cos b \sin c},$$

whence

$$\frac{1}{\cos q} = \sqrt{1 + \tan^2 q} = \frac{\sqrt{\cos^2 a + \cos^2 b - 2 \cos a \cos b \cos c}}{\cos b \sin c}.$$

It follows that

$$\cos x = \frac{1}{\sin c} \sqrt{\cos^2 a + \cos^2 b - 2 \cos a \cos b \cos c}.$$

5. A point P is located (Fig. 23) between two great circles AC and BC. The distances from P to these great circles are *p* **and** *q*, **respectively. The angle ACB between the great circles is C. What is the distance** *x* **from P to the intersection of the two given great circles?**

Let $ACP = y$, whence $BCP = C - y$. The two right-angled triangles give

$$\sin p = \sin x \sin y, \qquad \sin q = \sin x \sin (C - y).$$

Eliminate y. The first equation gives

$$\sin y = \frac{\sin p}{\sin x}, \quad \text{whence} \quad \cos y = \frac{\sqrt{\sin^2 x - \sin^2 p}}{\sin x}.$$

Substituting in the second equation, we get

$$\sin q = \sin C \sqrt{\sin^2 x - \sin^2 p} - \cos C \sin p.$$

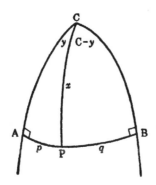

FIG. 23.—Point between two
great circles.

Transposing and squaring,

$$\sin^2 x - \sin^2 p = \frac{(\sin q + \cos C \sin p)^2}{\sin^2 C},$$

whence

$$\sin^2 x$$
$$= \frac{\sin^2 p \sin^2 C + \sin^2 q + \cos^2 C \sin^2 p + 2 \sin p \sin q \cos C}{\sin^2 C}$$

and, finally,

$$\sin x = \frac{1}{\sin C} \sqrt{\sin^2 p + \sin^2 q + 2 \sin p \sin q \cos C}.$$

Numerical Application.—What is the latitude of a point
in the Northern hemisphere, knowing its distances from the
zero meridian (45°) and from the 90°-meridian (30°)?

Let L be the latitude. The above formula becomes

$$\cos L = \sqrt{\sin^2 45° + \sin^2 30°},$$

$$\cos L = \frac{\sqrt{3}}{2},$$

whence

$$L = 30°.$$

Remark.—Note that, in this particular example, the point is located within a tri-rectangular triangle, so that (Problem 2)

$$\sin^2 45° + \sin^2 30° + \sin^2 L = 1.$$

6. In a spherical triangle (Fig. 24), the cosine of the arc which connects the middle points of two sides is proportional to the cosine of half the third side; the factor of proportionality is the cosine of one half of the spherical excess.

Required to prove:

$$\cos x = \cos E \cos \frac{2}{a}, \qquad \cos y = \cos E \cos \frac{b}{2},$$

$$\cos z = \cos E \cos \frac{c}{2}.$$

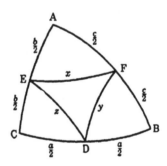

Fig. 24.—Arc that connects the mid-points of two sides.

Proof.—In the triangle AEF,

$$\cos x = \cos \frac{b}{2} \cos \frac{c}{2} + \sin \frac{b}{2} \sin \frac{c}{2} \cos A.$$

In the triangle ABC,

$$\cos a = \cos b \cos c + \sin b \sin c \cos A.$$

Taking cos A from each equation, and equating the two values, we get

$$\frac{\cos x - \cos \dfrac{b}{2} \cos \dfrac{c}{2}}{\sin \dfrac{b}{2} \sin \dfrac{c}{2}} = \frac{\cos a - \cos b \cos c}{4 \sin \dfrac{b}{2} \sin \dfrac{c}{2} \cos \dfrac{b}{2} \cos \dfrac{c}{2}}.$$

Hence

$$\cos x = \frac{\cos a - \left(2 \cos^2 \dfrac{b}{2} - 1 \right)\left(2 \cos^2 \dfrac{c}{2} - 1 \right) + 4 \cos^2 \dfrac{b}{2} \cos^2 \dfrac{c}{2}}{4 \cos \dfrac{b}{2} \cos \dfrac{c}{2}}$$

$$= \frac{\cos a + 2 \cos^2 \dfrac{b}{2} + 2 \cos^2 \dfrac{c}{2} - 1}{4 \cos \dfrac{b}{2} \cos \dfrac{c}{2}}$$

$$= \frac{\cos a + (1 + \cos b) + (1 + \cos c) - 1}{4 \cos \dfrac{b}{2} \cos \dfrac{c}{2}}$$

$$= \frac{1 + \cos a + \cos b + \cos c}{4 \cos \dfrac{b}{2} \cos \dfrac{c}{2}}.$$

This expression, divided by $\cos \dfrac{a}{2}$, is equal to cos E (Euler's formula). Hence

$$\cos x = \cos E \cos \frac{a}{2}.$$

Likewise for the other two formulae.

Second Proof.—In a plane triangle

$$c' = a' \cos B' + b' \cos A'.$$

This formula, applied to the derived triangle (Fig. 11), where $A' = 180° - A$, $B' = E$, $C' = A - E$, gives

$$\cos \frac{b}{2} \cos \frac{c}{2} = \cos \frac{a}{2} \cos E - \sin \frac{b}{2} \sin \frac{c}{2} \cos A$$

or

$$\cos E \cos \frac{a}{2} = \cos \frac{b}{2} \cos \frac{c}{2} + \sin \frac{b}{2} \sin \frac{c}{2} \cos A,$$

which is equal to $\cos x$, as seen in the triangle AEF (Fig. 24).

Remark.—The formula

$$\cos x = \cos E \cos \frac{a}{2}$$

can be derived directly in the following way.

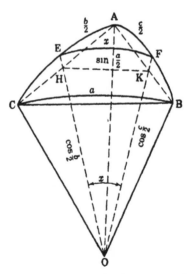

Fig. 25.—Geometrical construction
(perspective drawing).

Construction.—Join (Fig. 25) the vertices A, B, and C to the center O of the sphere of unit radius ($OA = OB = OC = 1$).

Join O to the middle points E and F of the sides b and c. Draw the chords AB, BC, and CA. Join H, the point of intersection of OE and CA, to K, the point of intersection of OF and AB.

Proof.—The formula giving a relation between the three sides and one of the angles of a plane triangle is applied to the triangle OHK.

Since the angle COA $= b$, we have, in the plane OCA,

$$OH = \cos \frac{b}{2}.$$

Likewise

$$OK = \cos \frac{c}{2}.$$

Since E is the middle point of the arc CA, H is the middle point of the chord CA. Likewise, K is the middle point of AB. Hence, in the triangle ABC, $HK = \dfrac{CB}{2} = \sin \dfrac{a}{2}$, for the chord subtending an arc is equal to twice the sine of half the arc $\left(CB = 2 \sin \dfrac{a}{2} \right)$.

Now the angle EOF is measured by the arc EF $= x$.

By plane trigonometry, we have, in the triangle OHK,

$$\sin^2 \frac{a}{2} = \cos^2 \frac{b}{2} + \cos^2 \frac{c}{2} - 2 \cos \frac{b}{2} \cos \frac{c}{2} \cos x,$$

whence

$$\cos x = \frac{2 \cos^2 \dfrac{b}{2} + 2 \cos^2 \dfrac{c}{2} - 2 \sin^2 \dfrac{a}{2}}{4 \cos \dfrac{b}{2} \cos \dfrac{c}{2}},$$

$$\cos x = \frac{(1 + \cos b) + (1 + \cos c) - (1 - \cos a)}{4 \cos \dfrac{a}{2} \cos \dfrac{b}{2} \cos \dfrac{c}{2} \cos \dfrac{a}{2}} \cos \frac{a}{2},$$

$$\cos x = \cos E \cos \frac{a}{2}.$$

7. The area of an equilateral triangle is S/n, where S designates the area of the sphere. Find the side _x_ of the triangle.

Let $A°$ be the angle of the triangle. By Girard's theorem (Appendix I),

$$\frac{1}{n} = \frac{\frac{1}{2}(3A - 180)}{360}, \quad \text{whence} \quad A = \frac{n + 4}{n} 60°.$$

The side x is given in terms of the angles by the fundamental formula (10)

$$\cos A = - \cos^2 A + \sin^2 A \cos x,$$

from which we deduce

$$\cos x = \frac{\cos A}{1 - \cos A} = \frac{\cos A}{2 \sin^2 \dfrac{A}{2}}$$

or, after substitution,

$$\cos x = \frac{\cos \dfrac{n + 4}{n} 60°}{2 \sin^2 \dfrac{n + 4}{n} 30°}.$$

Numerical Application.—Take $n = 4$. Show that $x = $ arc cos $(-\frac{1}{3})$.

$$\cos x = \frac{\cos 120°}{2 \sin^2 60°} = \frac{-\frac{1}{2}}{2 \left(\dfrac{\sqrt{3}}{2}\right)^2} = -\frac{1}{3}.$$

$$x = 180° - y, \quad \text{where} \quad \cos y = \frac{1}{3}.$$

$$\log \cos y = \overline{1}.522\ 8787$$
$$9002$$
$$\overline{215}$$
$$178\ 5$$
$$\overline{36\ 5}$$

$$y = 70°31'43.6''$$

$$x = 109°28'16.4''.$$

8. Given the dihedral angle (or edge angle) α between two slant faces of a regular *n*-sided pyramid, calculate the face angle *x* between two adjacent slant edges (Fig. 26).

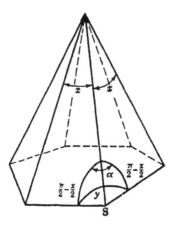

FIG. 26.—A regular pyramid.

The base of the pyramid is an *n*-sided regular polygon; the interior angle between two sides is equal to

$$y = \frac{n-2}{n} 180°.$$

Around the vertex S draw the unit sphere, so as to replace the trihedron S by a spherical triangle. Decomposing this isosceles triangle in two right-angled triangles, we get immediately

$$\sin \frac{y}{2} = \sin \frac{\alpha}{2} \cos \frac{x}{2},$$

whence

$$\cos \frac{x}{2} = \frac{\sin \dfrac{n-2}{n} 90°}{\sin \dfrac{\alpha}{2}}.$$

Discussion.—The problem is possible only if $0 < \cos \frac{x}{2} < 1$. $\left(\text{It is obvious that } \cos \frac{x}{2} \text{ cannot be negative, as } \frac{x}{2} \text{ must be smaller than } 90°.\right)$ The first condition demands

$$\sin \frac{n-2}{n} \, 90° > 0,$$

which implies $n \geqq 3$. $\left(\text{Note that } \sin \frac{\alpha}{2} \text{ is always positive.}\right)$ The second conditions requires

$$\sin \frac{\alpha}{2} > \sin \frac{n-2}{n} \, 90°,$$

that is to say $\alpha > \frac{n-2}{n} \, 180°$, which is evident (by geometry).

Numerical Application.—In a regular hexagonal pyramid, let $\alpha = 170°$. The general formula becomes

$$\cos \frac{x}{2} = \frac{\sin 60°}{\sin 85°} \quad \begin{array}{l} \ldots\ldots\ldots \ \overline{1}.937\ 5306 \\ \ldots\ldots\ldots \ \overline{1}.998\ 3442 \end{array}$$

$$\log \cos \frac{x}{2} = \overline{1}.939\ 1864$$

$$\begin{array}{r} 1953 \\ \hline 89 \\ 84 \\ \hline 5 \end{array}$$

$$\frac{x}{2} = 29°37'\ 7.4''$$

$x = 59°14'14.8''$.

9. Volume of a parallelopiped.

Let $OA = a$, $OB = b$, $OC = c$, be three edges making with one another (Fig. 27) the angles $BOC = \alpha$, $COA = \beta$, $AOB = \gamma$. Through C pass a plane normal to OA and containing CD, perpendicular to the plane AOB. The angle

CED = A is the dihedral angle of the edge OA. Draw the unit sphere around the vertex O.

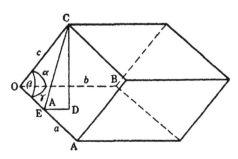

FIG. 27.—Volume of a parallelopiped.

$V = ab \sin \gamma . CD = ab \sin \gamma . CE \sin A = abc \sin \gamma . \sin \beta \sin A.$

But

$$\sin A = 2 \sin \frac{A}{2} \cos \frac{A}{2}$$

$$= \frac{2 \sqrt{\sin p \sin (p - \alpha) \sin (p - \beta) \sin (p - \gamma)}}{\sin \beta \sin \gamma},$$

in which $2p = \alpha + \beta + \gamma$, from the formulae (8).

It follows that

$$V = 2abc \sqrt{\sin p \sin (p - \alpha) \sin (p - \beta) \sin (p - \gamma)}.$$

Special Cases.

Volume of a rhombohedron ($b = c = a, \beta = \gamma = \alpha$):

$$V = 2a^3 \sqrt{\sin \frac{3\alpha}{2} \sin^3 \frac{\alpha}{2}}.$$

Volume of a right-angled parallelopiped ($\alpha = \beta = \gamma = 90°$):

$$V = abc.$$

Volume of a cube ($b = c = a, \alpha = \beta = \gamma = 90°$):

$$V = a^3.$$

CHAPTER VIII

EXERCISES[1]

1. The face angles of a trihedron measure 50°. Calculate one of the dihedral angles (edge angles).

2. In a triangle ABC, given $b = 45°$, $c = 60°$, $A = 30°$. Find B.

3. Two triangles ABC and BCD, both right-angled at C, have a side of the right angle in common, $BC = 59°36'30''$. The other side about the right angle is known in each case: $CA = 64°22'$, $CD = 52°5'$. Find the hypotenuses AB and BD.

4. Given, in a right-angled triangle $(C = 90°)$, $\cos c = 1/\sqrt{3}$ and $b = 45°$. Calculate the other two angles, each from the data only, and express the area of the triangle as a fraction of the area S of the sphere on which it is drawn.

5. Show that if $(C - E) : A : (B - E) = 2 : 3 : 4$, then

$$2 \cos \tfrac{1}{2}(C - E) = \sin \tfrac{1}{2}(b + c)/\sin \tfrac{1}{2}a.$$

[*Hint:* it is known that if the angles of a plane triangle A'B'C' are in the ratios $A' : B' : C' = 2 : 3 : 4$, then $2 \cos \tfrac{1}{2}A' = (a' + c')/b'$.]

6. An equilateral triangle ABC, with $a = b = c = 30°$, is drawn on a sphere of radius R. Find the area of the triangle expressed in dm^2 if $R = 6m$.

7. Let O, O', O'' be the middle points of the three edges of a cube intersecting in the same vertex C. Calculate the elements of the trihedron whose edges are OC, OO', OO''.

8. In an isosceles triangle, given $b = c = 32°$ and $B = 78°$. Find the side a.

9. In a regular tetragonal pyramid, the altitude is equal to one half the diagonal of the base. Calculate the dihedral angle between two adjacent slant faces.

10. Express $\sin (a + b)/\sin c$ in terms of the three angles of an oblique-angled triangle.

---

[1] Most of these exercises are questions that were asked by Cesàro at the entrance examination of the University of Liége.

11. The arc x that connects the middle points of two sides of an equilateral triangle of side a is given by $2 \sin \frac{1}{2}x = \tan \frac{1}{2}a$. (Compare Problem 6, Chapter VII.)

12. In a regular tetragonal pyramid, the angle of a face at the apex measures 60°. Calculate the dihedral angle between two adjacent slant faces.

13. In a regular pentagonal pyramid, the angle of a face at the apex measures 30°. Calculate the dihedral angle between two adjacent slant faces.

14. The altitude CH of a triangle ABC, right-angled at C, intersects the hypotenuse AB at H; arc $AH = b'$ and arc $BH = a'$. Prove that

$$\frac{\sin^2 a'}{\sin^2 a} + \frac{\sin^2 b'}{\sin^2 b} = 1.$$

15. Given a tri-rectangular trihedron O,XYZ. Plot $OP = 2$, $OQ = 3$, $OR = 4$, on the edges OX, OY, OZ, respectively. Calculate the elements of the trihedron P, OQR.

16. Let φ_1, ρ_1 be the longitude and polar distance (or colatitude) of a point P_1; φ_A, ρ_A, those of a point A. A 180°-rotation about A brings P_1 in P_2. What are the longitude φ_2 and the polar distance ρ_2 of P_2?

17. Let $AB = p$, $AC = q$, $AD = r$ be three edges of a right-angled parallelopiped. Calculate the elements of the trihedron whose edges are DA, DB, DC.—Numerical application: $p = 3$, $q = 4$, $r = 5$.

18. In a quadrantal triangle ($c = 90°$), given C and $2p$. Find (1) $\sin a \sin b$; (2) $\sin A \sin B$.

19. In a right-angled triangle (C = 90°), given the hypotenuse c and the spherical excess 2E. Find: (1) $\sin a \sin b$; (2) $\sin A \sin B$.

20. Given an equilateral triangle whose side is equal to 60°. Find the side of the triangle formed by erecting, at each vertex, a perpendicular to the corresponding bisector.

21. Solve a quadrantal triangle ($a = 90°$), given: (1) c, C; (2) A, $b + c$.

22. Prove the formulae of a quadrantal triangle ($a = 90°$):

$$\sin p = \cos \tfrac{1}{2}B \cos \tfrac{1}{2}C / \sin \tfrac{1}{2}A,$$
$$\cos p = \sin \tfrac{1}{2}B \sin \tfrac{1}{2}C / \sin \tfrac{1}{2}A,$$
$$\tan p = \cot \tfrac{1}{2}B \cot \tfrac{1}{2}C.$$

23. Consider a regular pentagonal prism with altitude h and the side of the base equal to a. Join O, one of the base vertices, to P and Q, the top vertices opposite the two adjacent vertical faces. Calculate the angle POQ.—Numerical application: $a = 4$, $h = 8$.

24. Same question for a hexagonal prism.—Numerical application: $a = 5$, $h = 5$.

25. Join a point P, inside an oblique-angled triangle ABC, to the vertices and produce the arcs AP, BP, CP to their intersections, D, E, F, with the sides of ABC. Prove that

$$\frac{\sin CE}{\sin EA} \cdot \frac{\sin AF}{\sin FB} \cdot \frac{\sin BD}{\sin DC} = 1.$$

26. In a tri-rectangular triangle ABC, join the middle points B' and C' of the sides AB and AC by the arc B'C'. Find the ratio of the areas of the triangles AB'C' and ABC.

27. In a triangle ABC, let α, β, γ be the lengths of the bisectors. Prove:

(1) $\cot \alpha \cos \tfrac{1}{2}A + \cot \beta \cos \tfrac{1}{2}B + \cot \gamma \cos \tfrac{1}{2}C$
$$= \cot a + \cot b + \cot c;$$

(2) $$\frac{\cot \alpha}{\cos \tfrac{1}{2}A(\cos B + \cos C)} = \frac{\cot \beta}{\cos \tfrac{1}{2}B(\cos C + \cos A)}$$
$$= \frac{\cot \gamma}{\cos \tfrac{1}{2}C(\cos A + \cos B)};$$

(3) $$\sin \alpha = \frac{2\sqrt{\sin b \sin c \sin p \sin (p - a)}}{\sqrt{\sin^2 b + \sin^2 c + 2 \sin b \sin c \cos a}}.$$

28. In a triangle ABC, given $a = b = 90°$ and $C = 60°$. Join any point P inside the triangle to the vertices A, B, C by arcs x, y, z. Find a relation between x, y, z.

29. In a quadrilateral ABCD, given: $B = C = D = 90°$, $BD = a$, and $AB = b$. Calculate the diagonal AC.

30. Show that the radii R_c, R_i, R_a of the circumscribing, inscribed, and escribed circles of a spherical triangle are given respectively by:

$$\tan^2 R_c = \frac{-\cos S}{\cos(S - A) \cos(S - B) \cos(S - C)},$$

where $2S = A + B + C$;

$$\tan^2 R_i = \frac{\sin(p - a)\,\sin(p - b)\,\sin(p - c)}{\sin p},$$

where $2p = a + b + c$;

$$\tan^2 R_a = \frac{\sin p\,\sin(p - b)\,\sin(p - c)}{\sin(p - a)},$$

where R_a is the radius of the escribed circle which touches a.

ANSWERS TO EXERCISES

Numerical applications have been calculated with five-place tables. The answer is followed, between parentheses, by the number of the formula to be applied.

1. $66°58'$ (8).

2. $B = 49° 6'24''$ (12).

3. $AB = 77°21'28''$, $BD = 71°53'14''$ (13).

4. $A = 60°$, $B = 45°$, $T = S/48$ (15 and 16, no tables of logarithms needed).

5. The desired relation is read off the triangle of elements (Fig. 9).

6. $E = 3.5°$, $T = 439.8\ dm^2$ (5a and Appendix 1).

7. $COO' = COO'' = 45°$, $O'OO'' = 60°$, edge angle $OC = 90°$, edge angle $OO' =$ edge angle $OO'' = 54°44'$ (13 and 18).

8. $a = 14°48'17''$ (16).

9. $109°28'$ (14).

10. $\sin(a + b)/\sin c = (\cos A + \cos B)/(1 - \cos C)$. Apply (3) (4).

11. Apply (5a).

12. $109°28'$ (14).

13. $113°46'$ (17).

14. Apply (15) twice. Use $\sin^2 x + \cos^2 x = 1$.

15. Edge angle $PO = 90°$, $OPR = \arctan 2 = 63°26'\ 5''$, $OPQ = \arctan (3/2) = 56°18'35''$, edge angle $PR = 59°11'33''$, edge angle $PQ = 67°24'41''$, $RPQ = 75°38'11''$. Apply (13) and (18), the latter twice.

16. $\cos \rho_2 = \cos 2\rho_A \cos \rho_1 + \sin 2\rho_A \sin \rho_1 \cos (\varphi_A - \varphi_1)$, $\sin (\varphi_2 - \varphi_A) = \sin \rho_1 \sin (\varphi_A - \varphi_1)/\sin \rho_2$. Apply Problem 4, Ch. VII and (7). (This question is encountered in crystallography.)

17. The spherical triangle ABC is right-angled at A. Its parts are: $a = 47°57'51''$, $b = 38°39'36''$, $c = 30°57'50''$, $B = 57°15'15''$, $C = 43°50'17''$. Apply (13) and (18). First, determine b and c by plane trigonometry.

18. $\sin a \sin b = - \sin 2p/(1 + \cos C)$, $\sin A \sin B = - \sin 2p$ $(1 - \cos C)$. Apply (8b), then combine with (9).

19. $\sin a \sin b = \sin 2E(1 + \cos c), \sin A \sin B = \sin 2E/(1 - \cos c)$.

Apply (11) and (8), or consider polar triangle of quadrantal triangle in preceding exercise.

20. $90°$ (8a, then 16; no tables of logarithms needed).

21. Given c, C.—Write $\sin b = \cos c/\cos C$, $\sin A = \sin C/\sin c$, $\sin B = \cot c/\cot C$. Apply (7) (9) (12).

Given A, $b + c$.—Express $\cos \frac{1}{2}(B - C)$ and $\cos \frac{1}{2}(B + C)$ in terms of $\frac{1}{2}(b + c)$ and $\frac{1}{2}A$ by means of (3) (4), hence B and C. Likewise express $\sin \frac{1}{2}(b - c)$ in terms of $\frac{1}{2}(B - C)$ and $\frac{1}{2}A$ by means of (3) or (4). Since $(b + c)$ is known, you get b and c.

22. Express the corresponding relations in the (right-angled) polar triangle of the given triangle. Apply the Law of Sines to Fig. 11.

23. $\sin \frac{1}{2}(POQ) = \sin 18° \sin c$, with $\tan c = 2(a/h) \cos 36°$. Apply (15). For $a = 4$ and $h = 8$, $POQ = 22°24'53''$.

24. Apply (7) or (15). For $a/h = 1$, $POQ = 51°19'4''$.

25. Apply (9). Note three pairs of equal angles having common vertex P. (This question is encountered in crystallography.)

26. 0.216 (14 or 7, and Girard's Theorem, Appendix 1).

27. (1) Apply (12).—(2) Apply (10) (9).—(3) Apply (12) twice.

28. $4(\cos^2 x + \cos^2 y - \cos x \cos y) = 3 \sin^2 z$. Apply (7) and $\sin^2 a + \cos^2 a = 1$.

29. Let $x = AC$. It is given by the biquadratic equation $\cos^4 x - \cos^2 a \cos^2 b \cos^2 x - \cos^2 a \cos^2 b \sin^2 b = 0$. Apply (13) three times and (15) twice.

30. Consider a circle inscribed in a spherical triangle. It is tangent to the sides. The spherical distances from a vertex to the nearest two points of tangency are equal. The radii to the points of tangency are perpendicular to the corresponding sides. Apply the formulae of right-angled triangles.

SPHERICAL AREAS

1. *Sphere.*—The area S of a sphere is equal to four times that of a great circle, that is to say $S = 4\pi R^2$, if R designates the radius.

2. *Lune.*—The area of a lune of angle A° is to the area of the sphere as A is to 360.

Consider a number of meridians; they intersect in the North and South Poles. Any two meridians bound a lune (for example, the meridians 75°W and 80°W). The angle of the lune is the angle between the two meridians (in this case, 5°); it is measured by the arc they intercept on the equator. Obviously lunes of equal angles have equal areas (for they can be made to coincide). The areas of lunes are proportional to their angles (this is proved in the same manner as the proportionality of angles at the center of a circle to the arcs they subtend), whence the desired proposition.

3. *Spherical Triangle.*—Girard's Theorem: The area of a spherical triangle is to the area of the sphere as half the spherical excess (expressed in degrees) is to 360.

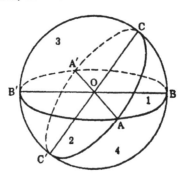

Fig. 28.—Area of a spherical triangle.

Consider a triangle ABC. Join its vertices A, B, C to the center O of the sphere (Fig. 28) and produce to the antipodes A', B', C'. Complete the great circles that form the sides of the triangle ABC.

The hemisphere in front of the great circle BCB'C' is thus divided into four triangles:

$$T_1 = ABC, \quad T_2 = AB'C', \quad T_3 = AB'C, \quad T_4 = ABC'.$$

From the construction it follows that the triangle A'BC is equal to the triangle $T_2 = AB'C'$. Hence, designating by lune A the area of a lune of angle A,

$$T_1 + T_2 = \text{lune A}.$$

Moreover

$$T_1 + T_3 = \text{lune B}$$

and

$$T_1 + T_4 = \text{lune C}.$$

On adding and transposing,

$$2\,T_1 = \text{lune A} + \text{lune B} + \text{lune C} - (T_1 + T_2 + T_3 + T_4),$$

whence

$$\frac{2\,T_1}{S} = \frac{A + B + C - 180}{360} = \frac{2\,E}{360}$$

and, finally,

$$\frac{T_1}{S} = \frac{E}{360}.$$

FORMULAE OF PLANE TRIGONOMETRY

DEFINITIONS: versine, coversine, exsecant, coexsecant, haversine.

$$\text{vers } x = 1 - \cos x, \qquad \text{exsec } x = \sec x - 1,$$
$$\text{covers } x = 1 - \sin x, \qquad \text{coexsec } x = \csc x - 1,$$
$$\text{hav } x = \tfrac{1}{2} \text{vers } x = \tfrac{1}{2}(1 - \cos x) = \sin^2 \tfrac{1}{2}x.$$

$$\sin 30° = \tfrac{1}{2}, \qquad \sin 45° = \frac{\sqrt{2}}{2}, \qquad \sin 60° = \frac{\sqrt{3}}{2}, \qquad \sin 90° = 1.$$

Complementary angles, x and $(90° - x)$, have complementary trigonometric functions, with the same sign:
$$\sin x = \cos (90° - x), \tan x = \cot (90° - x), \sec x = \csc (90° - x).$$

Supplementary angles, x and $(180° - x)$, have the same trigonometric functions, with opposite signs except *sine* and *cosecant*.

Antisupplementary angles, x and $(180° + x)$, have the same trigonometric functions, with opposite signs except *tangent* and *cotangent*.

Equal angles of opposite signs, x and $-x$, have the same trigonometric functions, with opposite signs except *cosine* and *secant*.

$$\sin x = \frac{\tan x}{\sqrt{1 + \tan^2 x}} = \frac{1}{\sqrt{1 + \cot^2 x}},$$

$$\cos x = \frac{1}{\sqrt{1 + \tan^2 x}} = \frac{\cot x}{\sqrt{1 + \cot^2 x}}.$$

$$\sin (x \pm y) = \sin x \cos y \pm \sin y \cos x,$$
$$\cos (x \pm y) = \cos x \cos y \mp \sin x \sin y.$$

$$\tan (x \pm y) = \frac{\tan x \pm \tan y}{1 \mp \tan x \tan y}.$$

$$\sin 2x = 2 \sin x \cos x, \qquad 2 \sin^2 \tfrac{1}{2}x = 1 - \cos x,$$
$$\cos 2x = \cos^2 x - \sin^2 x, \qquad 2 \cos^2 \tfrac{1}{2}x = 1 + \cos x,$$

$$\tan 2x = \frac{2 \tan x}{1 - \tan^2 x}, \qquad \tan^2 \tfrac{1}{2}x = \frac{1 - \cos x}{1 + \cos x}.$$

$$\frac{1 + \tan x}{1 - \tan x} = \tan (45° + x), \qquad \frac{\cot x + 1}{\cot x - 1} = \cot (45° - x).$$

$$2 \sin x \cos y = \sin (x + y) + \sin (x - y),$$
$$2 \cos x \cos y = \cos (x + y) + \cos (x - y),$$
$$2 \sin x \sin y = \cos (x - y) - \cos (x + y).$$

$$\sin P + \sin Q = 2 \sin \tfrac{1}{2}(P + Q) \cos \tfrac{1}{2}(P - Q),$$
$$\sin P - \sin Q = 2 \cos \tfrac{1}{2}(P + Q) \sin \tfrac{1}{2}(P - Q),$$
$$\cos P + \cos Q = 2 \cos \tfrac{1}{2}(P + Q) \cos \tfrac{1}{2}(P - Q),$$
$$\cos P - \cos Q = 2 \sin \tfrac{1}{2}(P + Q) \sin \tfrac{1}{2}(Q - P).$$

$$\frac{\cos P - \cos Q}{\cos P + \cos Q} = \tan \tfrac{1}{2}(P + Q) \tan \tfrac{1}{2}(Q - P).$$

SOLUTION OF TRIANGLES

LAW OF SINES: $\dfrac{a}{\sin A} = \dfrac{b}{\sin B} = \dfrac{c}{\sin C}.$

Corollary: $\cot B = \dfrac{c - b \cos A}{b \sin A}.$

LAW OF COSINES: $a^2 = b^2 + c^2 - 2bc \cos A.$

LAW OF TANGENTS: $\dfrac{\tan \tfrac{1}{2}(B - C)}{\tan \tfrac{1}{2}(B + C)} = \dfrac{\tan \tfrac{1}{2}(B - C)}{\cot \tfrac{1}{2}A} = \dfrac{b - c}{b + c}.$

Half-angle formulae:
$(2p = a + b + c)$
$$\begin{cases} \sin^2 \dfrac{A}{2} = \dfrac{(p - b)(p - c)}{bc}, \\[2mm] \cos^2 \dfrac{A}{2} = \dfrac{p(p - a)}{bc}, \\[2mm] \tan^2 \dfrac{A}{2} = \dfrac{(p - b)(p - c)}{p(p - a)}. \end{cases}$$

Area of the triangle: $T = \sqrt{p(p - a)(p - b)(p - c)}.$

INDEX